The arachnids: An introduction

The arachnids: An introduction

Keith R Snow

Lecturer in Zoology within proposed
North East London Polytechnic

New York:
Columbia University Press

Published in 1970
Columbia University Press
New York
© Keith R. Snow 1970
No part of this book may be reproduced
in any form without permission from
the publisher, except for the quotation
of brief passages in criticism

Library of Congress Catalogue Card Number 70-109151

SBN 231-03419-9

Printed in Great Britain

To my wife

Contents

Preface

With the reorganisation of many examination syllabuses the arachnids have, at long last, attained rightful recognition and are now included for study at an elementary level. Few texts are available dealing with this arthropod group and none is introductory in its nature while describing the animals in sufficient detail to enable the reader to appreciate fully their characteristics, appearances and ways of life.

The present text is concise and deals with the more common arachnids only. This plan has been adopted for several reasons: firstly, these are the animals selected for study by examination boards; secondly, the majority of them are readily seen and hence experience of them may be gained at first hand and, lastly, more is known about these forms than of the unincluded types and hence more complete accounts can be presented.

To supplement the text a number of diagrams have been included, all of which have been specially drawn by the author for this publication. The drawings are based on actual specimens and, except where indicated, no attempt has been made at schematisation.

It is hoped that this book will be useful to all students of zoology and natural history and that it will enhance the already growing interest in a group of invertebrates which have for too long been overshadowed by their insectan and crustacean relatives.

I should like to express my sincere thanks to Professor Don R. Arthur of King's College, London for reading the manuscript of this book and for his useful advice and criticism. My wife typed the manuscript in its final as well as in its numerous draft forms, and I am most grateful for her undertaking this onerous task.

London, 1969   K.R.S.

1

Introduction

Members of the phylum Arthropoda are bilaterally symmetrical, metamerically segmented, triploblastic coelomates which possess multi-jointed appendages on most of their segments. At least one pair of these appendages functions in feeding, the remainder being adapted for locomotion, holding the prey, reproduction, sense perception and other purposes. The cuticle of arthropods is composed of chitin together with a hardened protein called sclerotin, the latter giving the cuticle enormous strength and rigidity, and enabling it to function as a protective armour plating as well as conveying considerable powers of penetration to the mouthparts. In addition to these two functions, the impermeable cuticle minimises the problems of water-loss and osmoregulation, and acts as an external skeleton (exoskeleton), the body muscles taking their origin from inwardly directed cuticular processes. In order for movement of appendages and body regions to take place it is, of course, necessary for joints to intervene in the external skeleton. The joints are areas of soft cuticle called arthrodial membranes and occur between some or all segments, at intervals along the length of each segmental appendage, and within each segment between the individual cuticular plates. Basically, there are four plates per segment; a dorsal tergum, two lateral pleura (sing. pleuron) and a ventral sternum (Fig. 1).

As a result of the evolution of a hard body covering, the coelom has lost its importance in locomotion and is not employed to provide a hydrostatic skeleton as in members of the closely related phylum, the Annelida. The coelom has been reduced to elements associated with the reproductive and, in some cases, the excretory organs and a blood space or haemocoel has evolved as the main

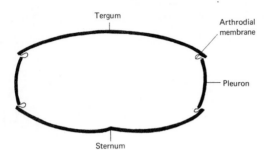

Figure 1 Transverse section through the body of an arthropod to show the arrangement of the segmental plates.

body cavity of arthropods. This is derived from remnants of the original coelom and from the embryonic blastocoele.

The Arthropoda is the largest of the animal phyla and contains about eighty percent of the animals in the animal kingdom. It is not surprising that within such a vast phylum several distinct lines of evolution can be detected. It is thus necessary to sub-divide the arthropods into a number of smaller categories called classes which reflect their phylogenetic origins. The arachnids constitute one of these classes, while the remaining divisions contain the insects, crustaceans, centipedes, millipedes, onychophorans, and the closely allied king-crabs.

It is essential for the arachnids to be distinguished from the remaining arthropod classes and for this purpose an outline classification of the phylum is given in Fig. 2, showing the main features of taxonomic importance.

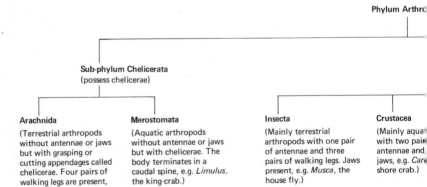

Phylum Arthro

Sub-phylum Chelicerata
(possess chelicerae)

Arachnida
(Terrestrial arthropods without antennae or jaws but with grasping or cutting appendages called chelicerae. Four pairs of walking legs are present, e.g. *Araneus*, the garden spider.)

Merostomata
(Aquatic arthropods without antennae or jaws but with chelicerae. The body terminates in a caudal spine, e.g. *Limulus*, the king-crab.)

Insecta
(Mainly terrestrial arthropods with one pair of antennae and three pairs of walking legs. Jaws present, e.g. *Musca*, the house fly.)

Crustacea
(Mainly aqua with two pai antennae and jaws, e.g. *Car* shore crab.)

Body organisation

The best way to understand the morphology and segmental
arrangement of any animal is to study the embryonic stages, for
during early development the segments are distinct and the appendages
of each segment are visible. The bodies of all embryo and most
adult arachnids can be divided into two distinct regions or tagmata
(sing. tagma): an anterior prosoma and a posterior opisthosoma
(Fig. 3). In the scorpions the latter region is again subdivided into
an anterior mesosoma and a posterior metasoma (Fig. 30). The
prosoma of all arachnids is composed of an anterior, unsegmented
region called the acron and six true segments. The reason for
considering the acron as an unsegmented region is that it does not
possess appendages, a coelomic cavity or a nerve ganglion in
either the embryo or adult stage. The appendages of the first
prosomatic segment are the chelicerae which, basically, are
composed of three sub-divisions or podomeres arranged to form
a pincer-like organ. In ticks and most spiders and mites the third
podomere has been lost and the chelate, grasping appendage is
converted into a piercing organ (Figs. 4 & 24). The pedipalps are
the appendages of segment two and are composed of a maximum
of six podomeres. These appendages serve as prehensile (grasping)
organs in scorpions, as sensory structures in ticks and mites, and
have the dual function of insemination and the sifting of food in
spiders.

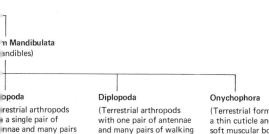

Figure 2 The main classes of arthropods and their characteristics.

... Mandibulata
(...andibles)

...opoda
(...restrial arthropods
... a single pair of
...nnae and many pairs
...gs. Segments not
...d in pairs, e.g.
...obius, a centipede.)

Diplopoda
(Terrestrial arthropods
with one pair of antennae
and many pairs of walking
legs. Segments fused in
pairs [diplosegments],
e.g. *Iulus*, a millipede.)

Onychophora
(Terrestrial forms with
a thin cuticle and a
soft muscular body wall,
e.g. *Peripatus*.)

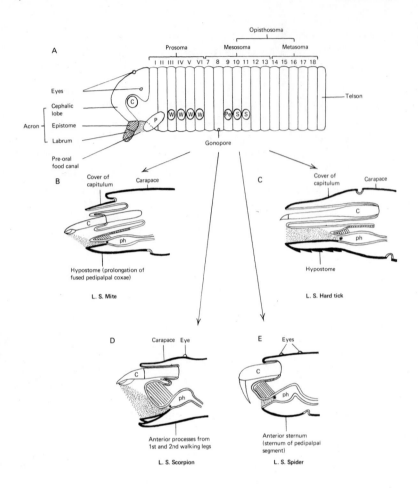

Figure 3 The segmentation of a generalised adult arachnid (A) and the ways in which the basic pattern of the anterior region is modified in mites, ticks, scorpions and spiders (B–E). It is not possible to show the fate of the cephalic lobe in this diagram but in all arachnids it develops into the region between the cheliceral bases extending from the epistome to the carapace, and the region of the carapace immediately surrounding the eyes.

The abbreviations used in this figure are as follows:
C, chelicera; P, pedipalp; W, walking leg; Pe, pectine of scorpion; S, spinneret of spider; ph, pharynx. In all of the diagrams the labrum, epistome and pre-oral food canal are similarly shaded, and the position of the mouth is indicated by an asterisk.

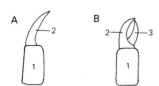

Figure 4 The arrangement of podomeres in the chelicera of A., a spider and B., a scorpion.

The appendages of segments three to six are the walking legs and each comprises seven podomeres to which specific names have been given. The basal division is called the coxa and this is followed by the trochanter, the femur, the patella, the tibia, the metatarsus and, finally, the tarsus which is equipped with two or three curved claws and, in some arachnids, a suctorial pad called the pulvillus. In addition to their normal function of locomotion the legs assist in the capture of food in spiders, while in ticks, mites and spiders one or all may bear special sensory organs.

Segments seven to eighteen (nineteen in scorpions) are collectively termed the opisthosoma and are followed by an unsegmented region, the telson. The seventh segment is not apparent in the adults of all arachnids but in spiders forms the waist or pedicel. The genital aperture or gonopore is located on segment eight and, in scorpions only, a pair of comb-like sensory structures (pectines) are found on segment nine. Spinnerets or silk-spinning organs are restricted to the spiders and occur on segments ten and eleven, there being no equivalent appendages in other arachnids. Lung-books are found in spiders and scorpions and open to the exterior in the anterior region of the opisthosoma. A maximum of two pairs are present in spiders while scorpions possess four pairs. Appendages are lacking from the remainder of the opisthosoma but in scorpions the post-segmental telson is modified to form a median, piercing organ called the sting.

The mouth of arachnids is not readily visible from the external surface, being located at the bottom of a tube, the pre-oral food canal, formed by the apposition of the labrum above, the pedipalpal coxae at the sides, and an under-lip which may be the ventral region of the pedipalpal segment (anterior sternum) as in spiders, forwardly directed processes from the bases of the first and second legs as in the case of scorpions, or a prolongation of the fused pedipalpal bases called the hypostome as in ticks and mites (Fig. 3). The remainder of the alimentary canal consists of a fore-gut (stomodaeum), a mid-gut (mesenteron) and a hind-gut (proctodaeum). Fig. 5 illustrates the alimentary systems of a tick and a spider as examples of a parasitic and a free-living arachnid.

Arachnids do not possess jaws (mandibles) but, as already mentioned, they have cutting or piercing appendages called chelicerae. It is thus not possible for arachnids to masticate their food using their mouthparts. Because of the basic lack of organs of mastication arachnids are, in the main, fluid feeders. After piercing the body wall of their prey with the chelicerae they either ingest the fluid contents or digest the tissues of their prey externally with enzyme-containing secretions which are discharged from the salivary glands (ticks and mites) or the mid-gut (spiders).

A powerful suctorial pharynx has evolved in arachnids for the purpose of ingesting the fluid meal. This draws the fluid up via the pre-oral food canal and pumps it into the mid-gut, which has a large capacity and is ideally suited for the reception of large quantities of liquid food. The ticks and some of the mites have become parasitic on vertebrate animals but their method of feeding is little modified from that of free-living arachnids for the former subsist on liquid blood, extravascular free tissue fluids and the products of digestion of tissue cells.

Gaseous exchange in arachnids occurs in a variety of ways. The respiratory gases may diffuse through the cuticle as in some mites and in larval ticks, or they may enter and leave the body through specialised structures which may be either lung-books or spiracles (Fig. 6). Lung-books consists of invaginations of cuticle which increase the surface area for the exchange of gases. The cuticle in such regions is extremely thin and thus allows the respiratory gases to pass through rapidly. It is conceivable that spiracles with their associated tracheal systems have evolved from lung-books by an inward extension of the cuticular folds so forming an internal network of tubes capable of conducting gases to and from the body tissues.

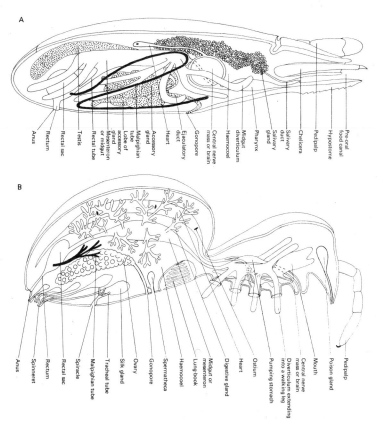

Figure 5 Lateral dissections of A, a male hard tick and B, a female spider to show the major internal organs.

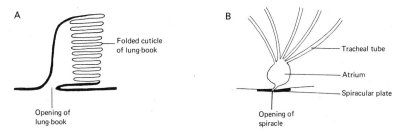

Figure 6 Schematic sections of A., a lung-book and B., a spiracle and associated tracheal tubes.

The circulatory system, like that of all arthropods, is described as 'open' for the arteries which convey blood from the heart open into an enlarged blood space. The heart is a tubular organ situated dorsal to the mid-gut and possesses a variable number of openings, or ostia, through which blood from the haemocoel is returned to the heart (Fig. 5).

The central nervous system, too, is similar to that found in other arthropods and basically consists of two cerebral ganglia united to a pair of sub-oesophageal ganglia by means of a circum-oesophageal commisure. In the more primitive arachnids, as in scorpions, the nerve cord bears a pair of nerve ganglia in many of the body segments, but in the majority of arachnids and, indeed, most members of other arthropodan groups, the segmental nerve cord ganglia have migrated forward and fused with the sub-oesophageal ganglia to form a compact mass of nervous tissue. This is seen clearly in mites, ticks and spiders. The central nervous system correlates and co-ordinates information gathered by the sense organs and initiates the appropriate responses of the effector organs, as well as being the site of implantation of instinctive or innate behavioral patterns. Arachnids possess a number of sense organs, many of which are associated with the outer body covering or cuticle. The most common cuticular sense organs are the hair-like setae which are receptive to touch, temperature, humidity and chemicals and which are distributed over the whole of the body surface of all arachnids. In scorpions and spiders there are long, fine sensory processes called trichobothria. These are located on the surfaces of the legs and pedipalps in spiders but in scorpions they are restricted to the pedipalps. Trichobothria are thought to be sensitive to air vibrations and may act as auditory organs. Complex sensory organs are present on the tarsi of all of the walking legs of spiders and on the tarsus of the first leg of ticks and mites. In spiders these are chemoreceptive and in ticks and mites they also perceive the stimuli of temperature and humidity. Most arachnids have, in addition to the sense organs already described, a number of photoreceptors or eyes, which may number as many as twelve in scorpions. On the other hand they may be absent as in some

species of ticks. Their distribution, too, is unusual being located
on the upper surface of the carapace.

It is difficult to generalise with regard to the reproductive systems
of arachnids owing to their considerable diversity. The male system
may consist of one or two compact testes or a single diffuse testis
and the spermatozoa produced are conveyed to a median gonopore
through one or two vasa deferentia. The method of insemination
of the females is also variable, for the spermatozoa may issue
from the male gonopore in a liquid medium, for example in spiders
and most mites, or they may be contained in packages called
spermatophores as in ticks and scorpions. In the harvestmen and
some mites an intromittent organ or penis is present to conduct
the spermatozoa into the female during mating but in other
arachnids no such organ is present and body organs not usually
associated with reproduction are commonly employed to assist
insemination. The female reproductive system is as variable as
that of the male and the single or paired ovary may be either
compact or diffuse and one or two oviducts lead to the median
gonopore. A further complication arises in female spiders in that
there are separate apertures for the entrance of the spermatozoa
and the exit of fertilised eggs, while scorpions are viviparous,
that is to say the females do not lay eggs but the young are born as
miniatures of the adults.

Excretion within the Arachnida is by means of Malpighian tubules
and coxal glands. The former are blindly ending tubes which arise
from the posterior region of the mesenteron and are reminiscent
of the similarly named organs of insects. There is, however,
evidence to suggest that the tubules have arisen independently in
the two arthropod classes and that they are, therefore, not
homologous. In arachnids the tubules remove waste products of
metabolism from the surrounding haemocoel and convert them into
guanine. This excretory substance is conducted along the tubules
to the hind-gut for subsequent elimination.

The paired coxal glands are located in the prosoma and are
composed of an excretory saccule positioned adjacent to the coxa
of the first walking leg, from which leads a coiled tubule, the

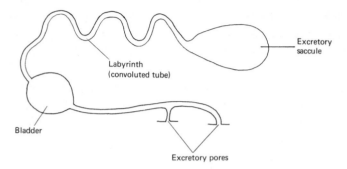

Figure 7 Diagrammatic representation of a coxal gland to show the basic features.

labyrinth. An enlargement or dilation of the labyrinth called the bladder follows and this leads to a straight tube which opens to the exterior behind the bases of the first and third walking legs (Fig. 7). Numerous variations on this basic plan are found within the arachnids and hence this description of a coxal gland can serve only as a guide. An organ of similar structure and function as the arachnid coxal gland is found in the class Crustacea, where it is called the 'green gland' or antennary gland. Once again this would appear to be an example of the independent adoption of a similar structure by separate groups of animals.

Arachnid classification
Each subdivision of the class Arachnida is given the status of an order and six of these contain common arachnids and hence are considered here. These orders are:

1 *Order Acarina* (ticks and mites)   Arachnids with no division between the prosoma and the opisthosoma, but with an anterior capitulum or false head bearing the mouthparts. In the vast majority of forms the chelicerae are non-chelate but in a few more primitive mites chelate chelicerae are found. Usually, there are no external signs of segmentation. The larval stages possess three pairs of walking legs in contrast to the normal arachnid number of four pairs found in the nymphs and adults. Members of the Acarina may be free-living, ectoparasitic or endoparasitic.

2 *Order Scorpiones* (scorpions)   Free-living arachnids in which
the dorsal surface of the prosoma is covered by a hard plate or
carapace, while the softer, externally segmented opisthosoma is
divided into two distinct regions – an anterior mesosoma and a
posterior metasoma armed with a terminal sting. Scorpions are
also characterised by their chelicerae and pedipalps, both of which
are chelate, and by the fact that they are viviparous.

3 *Order Araneida* (spiders)   Arachnids in which the prosoma is
covered by a dorsal plate or carapace and is separated from the
softer opisthosoma by a waist. Spiders possess spinning organs
(spinnerets) which are associated with silk-producing glands. The
chelicerae are not chelate and all members of the order are free-
living.

4 *Order Opiliones* (harvestmen)   Free-living, spider-like arachnids
which lack a waist and possess a segmented opisthosoma. The
chelicerae are chelate and the pedipalps are leg-like.

5 *Order Pseudoscorpiones* (false scorpions)   Small, free-living,
scorpion-like arachnids which lack a division of the body into a
mesosoma and a metasoma.

6 *Order Solifugae* (sun spiders)   Spider-like arachnids which do
not have the prosoma and opisthosoma divided by a waist. The
chelate chelicerae are large and the leg-like pedipalps bear suckers.
The body and appendages of these free-living arachnids are covered
by numerous, long setae.

Key to the main orders of arachnids
1 Pedipalps large and chelate   2
   Pedipalps either leg-like, or reduced in size or otherwise
   modified but never chelate   3
2 Opisthosoma divided into two distinct regions – a mesosoma
   and a metasoma   SCORPIONES
   Opisthosoma lacking a division   PSEUDOSCORPIONES
3 Segmentation of opisthosoma is visible externally   4
   Segmentation of opisthosoma is not visible externally   5

4 Leg-like pedipalps bear claws   OPILIONES
   Leg-like pedipalps bear suckers   SOLIFUGAE
5 Prosoma and opisthosoma separated by a waist   ARANEIDA
   No visible separation of prosoma and opisthosoma   ACARINA

2
Mites

The order Acarina contains a vast assemblage of forms which, collectively, are known as mites. One group of these acarines, the ticks, differs sufficiently from the remainder in their comparatively large size and their exclusive habit of parasitising vertebrate animals. Because of their morphological differences and their economic importance the ticks deserve special consideration and form the subject of Chapter 3. However, ticks are not the only group of mites of economic note and numerous other species are parasitic on man, his domestic stock, his crops and his stored foods. Of course, not all mites are parasitic, there being numerous species which have a free-living existence and they inhabit both terrestrial and aquatic environments.

Because of the variety of organisation in mites it is impossible to deal with them as a homogeneous group, and accordingly the order to which they belong is sub-divided into a number of smaller groups or sub-orders. These sub-divisions, which are separated largely on the position and number of spiracles, are listed below.

Sub-order 1 *Notostigmata*. A small group of primitive mites possessing four pairs of dorsal opisthosomatic spiracles, clawed pedipalps, and an externally segmented opisthosoma. These forms are brightly coloured and outwardly resemble harvestmen.

Sub-order 2 *Holothyroidea*. A group of primitive mites composed of a few species only, which are restricted to the Indo-Pacific region. Again they have clawed pedipalps but the external segmentation of the opisthosoma is not apparent as the whole of the dorsal and ventral surfaces are covered by large chitinous plates. Holothyroid mites possess two pairs of prosomatic spiracles.

Sub-order 3 *Mesostigmata*. Some members of this large group are parasitic and all are characterised by the presence of chitinous plates on the dorsal surface and a single pair of lateral spiracles in the region of the base of each third walking leg (Fig. 8B).

Sub-order 4 *Ixodoidea*. This is the group of large ectoparasitic mites known as ticks. They can be distinguished by the presence of a toothed lower lip or hypostome and, in nymphs and adults, by a single pair of spiracles located just posterior to the base of the third or fourth walking legs (Fig. 11). Spiracles are absent in larvae.

Sub-order 5 *Trombidiformes*. Many of the members of this group are parasitic on plants and animals, while others are free-living and predatory. Some free-living trombiculid mites have become secondarily aquatic and these are known as the Hydracarina or the water mites. All members of the sub-order have a single pair of spiracles which are found anteriorly in the region of the mouthparts (Fig. 8D, 8E).

Sub-order 6 *Sarcoptiformes*. This is an artificial group but is retained here as a matter of convention. It contains parasitic forms known as 'itch mites' which lack both spiracles and a tracheal system, and free-living mites or oribatids which possess four pairs of spiracles near the leg bases (Fig. 8A, 8C).

Sub-order 7 *Tetrapodili*. These mites are worm-like in their appearance and are equipped with two long setae which trail from the posterior end of the body. They do not possess spiracles and all stages have four legs only. Members of this group are all plant parasites and are known as 'gall mites'.

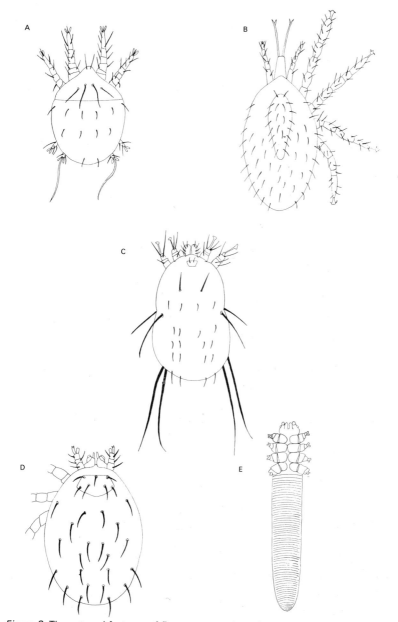

Figure 8 The external features of five representative mites.
A, *Acarus* (Sarcoptiformes), dorsal. B, *Ornithonyssus* (Mesostigmata), dorsal. C, *Sarcoptes* (Sarcoptiformes), dorsal. D, *Trombicula* larva (Trombidiformes), dorsal. E, *Demodex* (Trombidiformes), ventral.

External features

In the main, mites are small arachnids and, with the notable
exception of the ticks, usually measure less than 1 mm. in length.
Although a few primitive mites show signs of external segmentation
the majority do not, neither is the body divisible into an anterior
prosoma and a posterior opisthosoma. The dorsal surface of the
body may be covered by a single chitinous plate, or by several
smaller plates, or such plates may be lacking.

The chelicerae and pedipalps are located anteriorly on a subdivision of the body known as the capitulum. This region may be
freely exposed and hence the mouthparts are seen when the animal
is viewed from above, or else it may be retracted into an anterior
region of the post-capitular body (idiosoma) known as the
camerostomal fold. As in all arachnids a pre-oral food canal is
present and this is formed by either the labrum or chelicerae
dorsally, the pedipalpal coxae laterally and a forward extension of
the bases of the pedipalpal limbs (hypostome) ventrally.

The chelicerae are composed of either two or three podomeres
and may be either chelate as in scavenging or predatory forms
where they are used for prehension, or stylet-like or armed with
movable cutting digits as in parasitic forms (Fig. 9). Variation is
seen in the pedipalps also and these may be either leg-like, or
large and clawed, or small with a few podomeres only and function
as sensory organs.

In the majority of forms the legs are well developed and each
tarsus is equipped with a pair of curved claws. In aquatic mites,
however, the legs have lost their ambulatory function and their
surfaces are covered with numerous, long setae which increases
their surface area and hence makes them suitable for use as
paddles. Yet another modification is seen in some mites where the
hind legs are elongate and are used for jumping.

Of the remaining body openings of mites, the anus is located
near to the posterior extremity of the body, but the position of the
spiracles and gonopore are variable. The spiracular openings also
vary in number as indicated earlier in the chapter and the gonopore
may be positioned as far forward as the level of the anterior walking

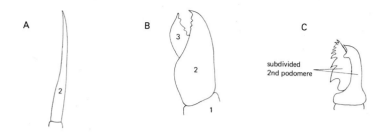

Figure 9 Various types of chelicerae found within the order Acarina.
A, stylet-like as in *Dermanyssus* (Mesostigmata), B, chelate as in *Opilioacarus*
(Notostigmata) and C, denticulate as in *Ixodes* (Ixodoidea).

legs although it may be found in the posterior half of the body.

Because of its larger size the tick has been chosen as the
representative of the Acarina for detailed study and is described in
the next chapter.

Anatomy, physiology and life histories
Mites are fluid-feeders and obtain their nutriment from living or
dead animals and plants, the food being detected by the general
body setae, the specialised setae of the first walking leg, and in
trombiculids, oribatids and some ticks, additionally by the eyes,
although this last suggestion is disputable. Thus some are parasites,
some predators and some scavengers but in all cases a correlation
is seen in the diet and the form of the chelicerae (see above). In
some forms external digestion occurs under the influence of a
secretion from the paired salivary glands and in all mites the fluids
enter the pre-oral food canal by the suctorial action of the pharynx
and pass along the oesophagus to the mid-gut which is extensive
owing to the presence of a number of diverticula. In the trombiculid
mites the mid-gut ends blindly but in other forms it leads via a
short rectum to the anus. At the boundary of the mid-gut and the
hind-gut a single pair of Malpighian tubules can be seen in all mites
except the trombiculids where the excretory tubules open directly at
the 'anus'. In addition to the Malpighian tubules some mites

possess one to four pairs of coxal glands and these function in excretion and osmoregulation.

Respiration in some small adult mites and in most larval stages is by diffusion through the body cuticle, but in larger forms a tracheal system has evolved which communicates with the exterior by means of the one to four pairs of spiracles. With the tissues being well ventilated the circulatory system plays but a minor role in the translocation of respiratory gases and the heart is, in fact, very small and may be absent altogether in some forms.

The male reproductive system is composed of a pair of compact testes with a vas deferens leading from each into a single accessory gland. A short ejaculatory duct leads from this gland, its opening to the exterior being termed the gonopore. In the males of some species an intromittent organ is present to transmit the spermatozoa into the female, while in other males, there is no such organ and the spermatozoa may be packaged in the form of a spermatophore.

Female mites may have one or two oviducts but only a single ovary. The oviducts lead into a seminal receptacle and an accessory gland, and a short vagina leads to the exterior to open at the median gonopore.

In mesostigmatid mites mating entails the male and female positioning themselves so that their gonopores are in apposition. The intromittent organ of the male is then introduced into the female and spermatozoa are subsequently transferred. Other mites, for example ticks, produce spermatophores and hence many spermatozoa are introduced in a single 'package'. An intermediate condition between insemination by free spermatozoa and spermatophores is seen in some oribatids where the male deposits a stalked cup on the ground and ejects spermatozoa into it. When a female locates this cup she positions herself over it and draws the spermatozoa into her genital tract. When the eggs of mites are fertilised they are usually laid in cracks and crevices in the soil although in some parasitic forms they are laid on or in the skin or in the respiratory tract of their host.

Although a few oribatids are ovoviviparous, the majority of mites are oviparous, and in both cases the eggs hatch into six-legged larvae which normally require to feed before moulting to produce the first nymphal instar. Some mites have one or two additional nymphal stages in which case the terms protonymph, deutonymph and tritonymph are employed to describe them. All of the nymphal instars require to feed in order to moult and their final ecdysis produces the adult mite. The nymphs resemble the adults in possessing eight legs but may be differentiated from them by their smaller size and the absence of a gonopore.

As well as showing great diversity in their morphology, mites also display enormous variety in their life-histories. Many are free-living, for example, the 'grain mite' *Acarus*, while others have become ectoparasitic by feeding on the external surfaces of vertebrate animals. Yet others have entered the respiratory tracts or bored into the skin of vertebrates and so have adopted an endoparasitic mode of life.

Trombicula, the 'chigger mite', has a life cycle which cuts across the categories of life-histories given above for it is parasitic in the larval stage, yet free-living in the later stages. The larvae attach to the skin of vertebrates, inject a salivary secretion which causes lysis of the tissues of the skin, and then feed on the liquefied tissues. Feeding lasts for between a few days and a month and when engorged they drop to the ground and moult to form non-parasitic protonymph instars. This is followed by a deutonymph, a tritonymph and, finally, an adult, all of which are free-living and feed on the eggs of insects. The adult female lays her eggs on the ground and these eventually hatch into parasitic hexapod larvae.

Trombicula is important because of its effect on man and his domestic animals. The secretion which causes lysis of the skin produces severe irritation and, in some species, contains rickettsial organisms which are transferred to the host as the larva feeds. Hence these mites are capable of transmitting serious diseases and so affect man's health and economy.

Another mite which causes severe skin irritation is the 'itch mite', *Sarcoptes*. The adult mites are found on the surface of the skin at night, maintaining their hold by means of stalked suckers found on the tips of the first and second pairs of legs in females and the first, second and fourth pairs in males. Additionally, there are a number of long setae on the remaining legs which aid their progression over the surface of the host. While the mites are on the surface they can be transferred to a new host by the contact of one host with another.

The females are fertilised whilst on the skin and subsequently burrow with the chelicerae, laying their eggs as they excavate more deeply until a chain of thirty to fifty eggs have been deposited. After a few days the hexapod larvae hatch and remain in the burrow to moult to the protonymph. This instar burrows off the main channel and soon moults to give the deutonymph and, later, the adult stage.

The 'chigger mite' and the 'itch mite' show only a few parasitic adaptations and these are concerned with skin penetration, feeding, and progression on the skin of the host. The 'hair follicle mite', *Demodex*, however, is hardly recognisable as a mite as it is worm-like and, therefore, well adapted for life in hair follicles. It possesses an expanded capitular region, four pairs of very short legs and an elongate body. The life-history is, however, typical in that there is a larval stage, three nymphal phases and the adult instar. There are a number of species of *Demodex*, one of which is found in man. This parasite does not produce symptoms in humans and is common in people with normal, healthy skins. Transmission of this mite, like *Sarcoptes*, is by contact.

3

Ticks

Now that the economic destruction by ticks has been appreciated these arachnids are regarded as one of the most serious pests known to man. Apart from the physical damage inflicted when they bite, namely the production of sores and ulcers, there are harmful effects caused by pathogenic organisms injected into hosts as the ticks feed, and these may produce severe illness and even death.

There are two distinct types of ticks, the hard ticks or ixodid ticks and the soft ticks or argasid ticks. Although possessing similar morphological features and almost identical anatomical characters, members of the two groups show considerable variation in some physiological phenomena and in their life-histories. These similarities and differences are considered below.

External features

1 *Ixodid ticks*
The unfed animal is flattened dorso-ventrally and possesses an anterior capitulum which bears the mouthparts, and a posterior region, the idiosoma, which bears four pairs of walking legs in all stages except the larva where there are only three (see Figs. 10 and 11).

Most of the dorsal surface of the idiosoma in the male is covered by a hard sclerotised plate called the carapace (or scutum), while in the female, nymph and larva the anterior part of the idiosoma only is so covered, the posterior region consisting of a soft, flexible cuticle. Eyes may be present on the margins of the carapace in some species but may be absent in others.

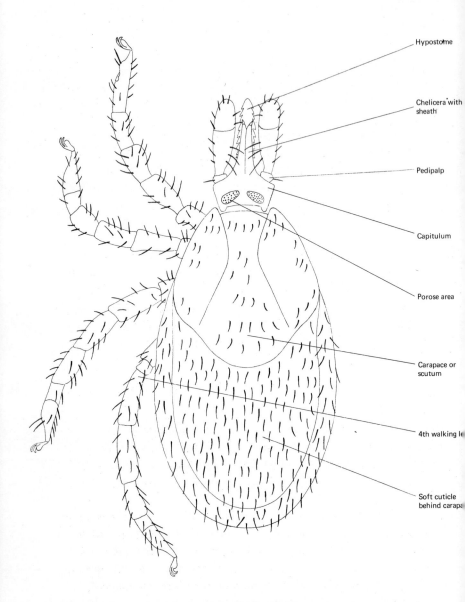

Figure 10 Dorsal view of a female ixodid (hard) tick.

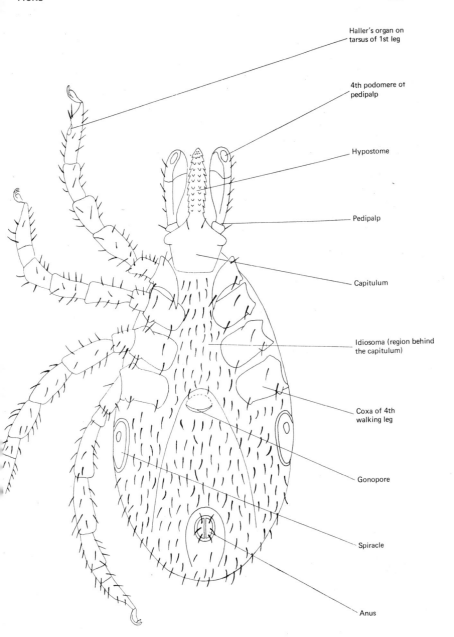

Haller's organ on
tarsus of 1st leg

4th podomere of
pedipalp

Hypostome

Pedipalp

Capitulum

Idiosoma (region behind
the capitulum)

Coxa of 4th
walking leg

Gonopore

Spiracle

Anus

Figure 11 Ventral view of a female ixodid (hard) tick.

The capitulum is composed of two segments; the appendages of the first are the chelicerae and these consist of two podomeres, an elongate shaft and a terminal cutting digit. When retracted, each chelicera is enclosed by a membrane called the cheliceral sheath, which is actually a continuation of the chelicera. Protraction of a chelicera entails the unfolding of this sheath (Fig. 12).

The appendages of the second segment, the sensory pedipalps, are made up of four podomeres and lie lateral to the chelicerae. Only three of these podomeres are visible from the dorsal surface, the fourth being housed in a ventral depression of the third podomere. A spatulate hypostome, bearing longitudinal rows of teeth or denticles on its ventral surface, is formed by the elongation of the pedipalpal bases and is ventrally placed to form a lower lip or channel along which nutrients and secretions pass. The capitulum of the female bears two densely pitted regions called the porose areas which produce a fluid to assist the movement of Gené's organ by lubricating it (see p. 32).

The remaining four appendages are the walking legs. The distal podomere (tarsus) of the first walking leg possesses a depression in its surface which contains a sense organ caller Haller's organ, the significance of which is discussed later. Each limb terminates in two curved claws and an elongate suction pad.

In the adult there are four remaining body apertures. They are the paired spiracles (openings of the respiratory system) which are located ventro-laterally just posterior to the coxae of the fourth walking legs; the genital aperture or gonopore found in the mid-line

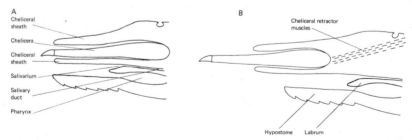

Figure 12 Longitudinal sections through the capitulum of a hard tick showing the chelicerae retracted (A) and protracted (B).

on the ventral surface and the anus, also in the ventral mid-line
and close to the posterior margin of the body. Of these four
structures only the anus is found in the larva, and the spiracles
are found in addition to the anus in the nymph. Owing to the
absence of a visible respiratory system in larvae it is assumed that
respiration takes place through the body cuticle.

In the males of some genera there are one to three pairs of hard
plates situated in the vicinity of the anus. The function of these
plates is unknown.

2 *Argasid ticks*

Although the basic structure of the argasid or soft ticks resembles
that of hard ticks there are several important differences between
the two groups. As the common name of the argasids suggests
they do not possess a dorsal carapace or scutum and hence it is
not easy to distinguish between males and females (Figs. 13 and
14). The cuticle is relatively soft and leathery and possesses
numerous folds and small plates called discs and mammillae. In
the larvae of some species, however, there is a small sclerotised
dorsal plate which is thought to be homologous with the carapace
of ixodids (Fig. 15). In fact, larval argasids resemble larval ixodids
in a number of respects. Thus the capitulum is similar except that
the four podomeres of the pedipalps are of about equal size and
that the fourth podomere is not located in a depression of the
third podomere.

Except for size, the first to fourth nymphal stages and the adult
are similar to one another; the nymphal stages are progressively
larger and the adults are the largest of all, with the females
invariably being larger than the males. In these forms the body
is flat and conceals the capitulum and most of the podomeres
of the legs. The capitulum is actually located in a depression in
the overlying cuticle called the camerostome and, except for the
pedipalps which resemble those of the larvae, it is similar to
that of the ixodids. The spiracles, genital aperture and the anus
are similarly placed to those of ixodids.

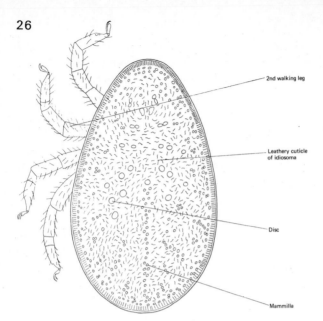

2nd walking leg

Leathery cuticle
of idiosoma

Disc

Mammilla

Figure 13 Dorsal view of an argasid (soft) tick.

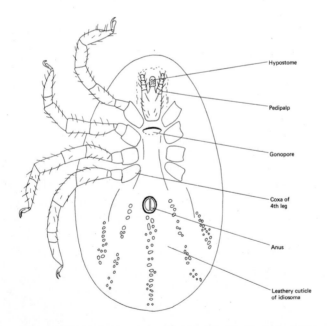

Hypostome

Pedipalp

Gonopore

Coxa of
4th leg

Anus

Leathery cuticle
of idiosoma

Figure 14 Ventral view of an argasid (soft) tick. (Cuticular folds omitted.)

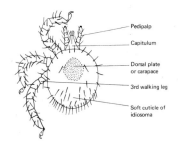

Pedipalp

Capitulum

Dorsal plate
or carapace

3rd walking leg

Soft cuticle of
idiosoma

Figure 15 Dorsal view of the larval
stage of a soft tick

Anatomy, physiology and life histories

Blood-sucking arthropods have been classified into two groups
according to their method of obtaining the blood component of
their meal. They may either (i) insert their mouthparts into the
lumen of a blood capillary and so feed on blood alone or (ii) insert
their mouthparts indiscriminately into the skin surface, causing
lysis of the tissues, and feed on the tissue and extracellular
fluids until they eventually break down the walls of blood capillaries
and produce a blood pool from which they then feed. These feeding
methods have been referred to as solenophagic and telmophagic
respectively. Ticks are telmophagic feeders.

Before feeding can commence the tick has to locate its host.
For this purpose the animal is equipped with various sense organs:
setae on the fourth pedipalpal podomere which are chemoreceptive;
tactile and thermoreceptive setae on the legs; and the special organ
on the tarsus of the first leg (Haller's organ) which is a humidity
and olfactory receptor. These link with a fused ganglionic mass
referred to as the 'brain' or central nerve mass which, as can be
seen from Fig. 5, is penetrated by the oesophagus.

The tick then climbs on to the host and moves over its body
surface, holding on by means of its tarsal claws. Eventually it makes
an incision into the skin with the cheliceral digits and the chelicerae
and hypostome are inserted into the wound. Digital shearing
continues until the mouthparts are fully embedded in the skin.
It is important to realise that the barbed hypostome plays no part
in skin penetration but prevents the animal from being dislodged
from the skin. In this function the hypostome may well be aided

by the numerous, small projections on the surfaces of the cheliceral sheaths.

In ixodid ticks the salivary glands then produce a secretion which sets to a latex-like consistency around the mouthparts and which functions to 'cement' the tick into the skin. This is another adaptation to prevent the tick being removed from the skin during its feeding period, which in ixodids may be as long as two weeks. In general, argasid ticks are fast feeders and no such secretion is produced. Common to both hard and soft ticks is the production of a histolytic secretion which liquefies the tissue below the mouthparts of the tick, the liquid ingested by the tick being sucked into the gut by the action of the muscular pharynx. Eventually the enzymic secretion breaks down the walls of the dermal blood vessels and the released blood is ingested.

The alimentary canal is well suited to an ectoparasitic mode of life. The ducts from the paired salivary glands open into the salivarium (Fig. 12) and their secretion, which contains anticoagulant and histolytic components, is conveyed into the skin by way of a pre-oral food canal formed by the apposition of the chelicerae, the hypostome and the pedipalps. This is also the route followed by the liquefied tissues and blood as a result of sucking by the pharynx which presumably alternates with the secretory process. From the pharynx the fluids pass along the oesophagus to the extensive mid-gut which consists of a central region from which radiates a number of caeca or diverticula (Fig. 5). This region is adapted for the reception of large quantities of liquid food material, and it is here that digestion and absorption take place.

The products of digestion are translocated around the body of the tick by means of the open circulatory system. This is simple in design and comprises a dorsal heart with a single artery leading from it anteriorly. Blood enters the heart from the haemocoel through a single pair of ostia and is pumped into the anterior blood vessel and thence into a sinus which surrounds the brain, oesophagus and nerves to the legs. Blood leaves the sinus, enters the haemocoel and eventually returns to the heart by way of the ostia.

As the food is utilised for synthesis and the production of energy so the waste products which result are excreted. In nymphs and adults carbon dioxide is exchanged for oxygen at the tracheoles, which are the fine terminal elements of the tracheal tubes, while in larvae the gases diffuse through the tissues to and from the body cuticle. Waste in the form of the solid nitrogenous product, guanine, is removed by the Malpighian tubules to the rectal sac for temporary storage before being voided.

It has been inferred above that a large quantity of food is ingested, and this is especially true for ixodid ticks. Thus larvae of hard ticks increase in weight by ten to twenty times, nymphs by forty to eighty times, males by one to three times and females by eighty to one hundred and twenty times; this would mean that an unfed female weighing about six milligrams would weigh over five hundred milligrams when fully fed. As might be expected, ixodids take a long time to imbibe these large quantities of fluid. Larvae normally require between two to four days to complete feeding, while nymphs take three to five days and females seven to twelve days. Some species of males do not feed, but those which do are intermittent feeders and feed over a period of a few days.

The rate of feeding in ixodids is not constant throughout the period of attachment, but can be divided into two distinct phases. The first period is characterised by slow feeding and lasts until twelve to twenty-four hours before full engorgement. During this phase the cuticle of the tick is being formed by the addition of an endocuticle, and during the second, short phase, which is characterised by rapid feeding, the newly formed endocuticle stretches as do the other cuticular components (Fig. 16). The body

Figure 16 Sections of the cuticle of a hard tick (diagrammatic) showing the appearance in the unfed animal (A), the growth of endocuticle during feeding (B) and the subsequent stretching of the endocuticle (C).

is thus much enlarged and the gut diverticula distend with fluids and fill the available space. Owing to their high degree of sclerotisation the capitulum, legs and carapace do not stretch, and this explains the small increase in weight shown in males, for their bodies bear extensive, hard plates and any distension which occurs is limited to the straightening out of folds of soft cuticle between these plates.

Ticks are in close contact with the warm body of their host throughout the period of attachment and will lose water by evaporation from the cuticle and spiracles. Since ixodid ticks remain attached for long periods much water is lost, but this is replaced by the water in the meal which they are imbibing and in this way they avoid desiccation. The removal of water is of value in that it concentrates the essential nutrients in the meal prior to their digestion, thus allowing additional food to be imbibed.

The short feeding time required for argasid nymphs and adults to become replete contrasts with the long feeds of ixodid ticks. This is because the cuticle of the rapid feeders possesses a fully formed endocuticle at the commencement of feeding and so the cuticle is immediately ready to distend. The time taken by argasid nymphs and adults to complete feeding rarely exceeds two hours and may be as short as five minutes. The short duration of attachment is paralleled by the comparatively small quantity of nutrients imbibed, which is about two to four times the unfed weight of the individual. Argasid larvae, unlike the nymphs and adults, require five to thirty days to become fully fed and simulate hard ticks in that they commence feeding with an incompletely formed cuticle. Typical weight increases for argasid larvae are in the region of fifteen to twenty-five times.

Argasid nymphs and adults are in contact with their hosts for a short time and consequently lose only a small quantity of water by evaporation. It is, therefore, not possible for them to concentrate their meal by this method and so an alternative mechanism has been evolved. This depends on the presence of a pair of osmoregulatory coxal glands which open to the exterior just behind each first coxa. They produce a copious secretion of water while the animal is feeding and so compensate for the limited effects of evaporation. Coxal glands are not present in ixodid ticks, thus emphasising their role in argasids.

Connected with the differences in feeding habits of the two tick groups are the differences in life-histories. Thus, although ixodid ticks imbibe large quantities of fluids at each feed, each instar feeds only once. There are, therefore, three feeding stages in a complete life cycle, and following her single feed the female lays one thousand to ten thousand eggs in a single batch. In contrast to this, soft ticks possess a larval stage which feeds once, two to four nymphal stages which may feed from two to seven times, and an adult stage which feeds about seven times, after each of which the female deposits a batch of fifty to one hundred and fifty eggs.

Because of the numerous hazards associated with a parasitic way of life it is essential that the females lay large numbers of eggs. This ensures that sufficient individuals reach the adult stage to perpetuate the species. In order to lay large numbers of eggs the female must imbibe sufficient nutrients to permit their development. This accounts for the large volume of nutrients imbibed by ixodid females, and the smaller but repeated feeds of the argasid females.

Before eggs can be laid the female must be fertilised and this occurs with the aid of a body structure not normally associated with insemination. Copulation entails the male climbing beneath the female either when she is feeding or before she commences feeding, depending on the species, and repeatedly inserting his mouthparts into her gonopore, an action which presumably enlarges the aperture. As the male withdraws his mouthparts a spermatophore issues from his gonopore and adheres to the ventral surface of the female. The male pushes it into the female gonopore by means of his hypostome and within the female genital tract the spermatozoa which are contained with the spermatophore are liberated.

The spermatophores are formed within a region of the male genital system called the accessory gland. This receives spermatozoa from the paired testes and discharges the completed spermatophores along a median ejaculatory duct. The female receives the large spermatophore in a sac-like structure called the seminal receptacle, and from this region the liberated spermatozoa swim up the oviduct to fertilise the eggs still within the single, horse-shoe shaped ovary.

Female ticks are unique among animals in possessing a glandular organ, Gené's organ, which is situated in a depression in the dorsal surface of the tick just posterior to the capitulum. As each egg emerges from the gonopore the organ of Gené is everted, and extends forward over the capitulum to the gonopore. It grasps each egg with its finger-like lobes and in so doing coats it with a sticky wax. This secretion minimises water-loss and enables the eggs to adhere to one another in a compact mass.

Larvae emerge from the eggs, seek a host, feed to engorgement, and eventually moult to produce the next instar. In argasids and most ixodids each instar feeds on a different host, and thus in the case of hard ticks three distinct hosts are utilized during the life cycle. Some ixodids, however, use only two hosts during a life cycle, with the larvae feeding and moulting on the host with the larval mouthparts still attached to the skin. The nymph subsequently emerges from the larval cuticle and reattaches to the same host. When fully fed the nymph detaches from the host, falls to the ground, moults and the adult seeks a new host. This pathway of evolution has been taken one stage further in some hard ticks where only one host is required. Hence the larva, nymph and adult feed on the same host, the larva and nymph moulting on the host after engorgement. The female leaves the host and in common with all ticks lays her eggs in cracks and crevices in the ground or in walls of buildings.

Ticks are transmitters of disease to man and his domestic animals and hence affect man's well-being and economy. Within recent years they have been implicated as vectors of many virus, spirochaete, bacterial, rickettsial and protozoan diseases, and to a large extent this accounts for the present-day interest in the group.

An important cattle disease transmitted by ticks in Britain is 'redwater fever', so called because the causal agent breaks down the red corpuscles and releases haemoglobin. This is excreted by the kidneys and colours the urine red. This protozoan parasite is transmitted by the hard tick, *Ixodes*.

4

Spiders

Spiders are among the best known of the terrestrial arthropods and, although the vast majority are completely harmless to man, they rank among the most feared of invertebrates.

The order Araneida, to which all true spiders belong, is divided into two sub-orders: the Araneomorphae, containing the more highly evolved species, and the Mygalomorphae with only a small number of representatives and with but one British example, *Atypus*.

External features

1 *Araneomorph spiders*

The body is readily divisible into an anterior prosoma and a posterior opisthosoma (Fig. 17). A plate or carapace covers the entire dorsal surface of the prosoma so that when seen from above it shows no visible signs of segmentation. A number of depressions in the surface of the carapace indicate the regions of attachment of internal muscles. The depression in the centre of the carapace is called the fovea and eight furrows or striae radiate from it in the direction of the walking legs. Anteriorly, the carapace bears a number of eyes. In most spiders eight eyes are present, arranged in two rows of four, the anterior median two differing from the remainder in possessing a direct retina and dark colouration in contrast with the indirect retina found in the others which has light reflected on to it by means of an internal process or tapetum (Fig. 26). The ventral surface of the prosoma is formed from two plates: a small anterior sternum, which is often misleadingly referred to as the labium, and a larger posterior sternum (Fig. 18).

33

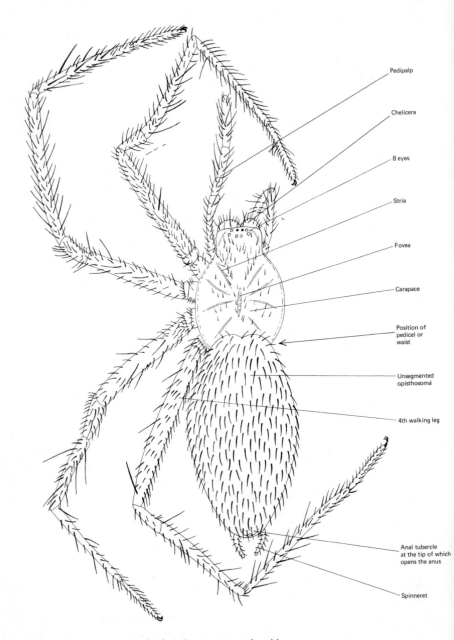

Pedipalp

Chelicera

8 eyes

Stria

Fovea

Carapace

Position of
pedicel or
waist

Unsegmented
opisthosomá

4th walking leg

Anal tubercle
at the tip of which
opens the anus

Spinneret

Figure 17 Dorsal view of a female araneomorph spider.

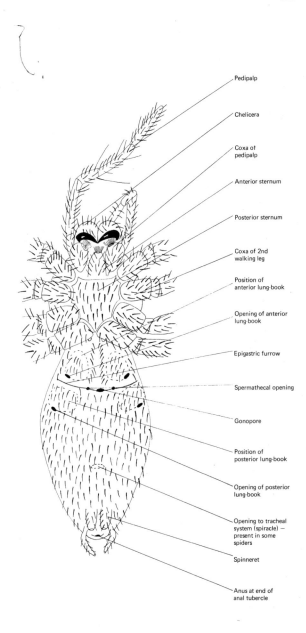

Figure 18 Ventral view of a female araneomorph spider.

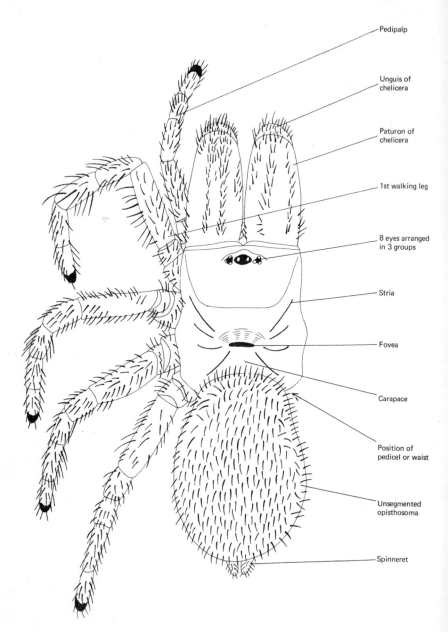

Pedipalp

Unguis of
chelicera

Paturon of
chelicera

1st walking leg

8 eyes arranged
in 3 groups

Stria

Fovea

Carapace

Position of
pedicel or waist

Unsegmented
opisthosoma

Spinneret

Figure 19 Dorsal view of a mygalomorph spider.

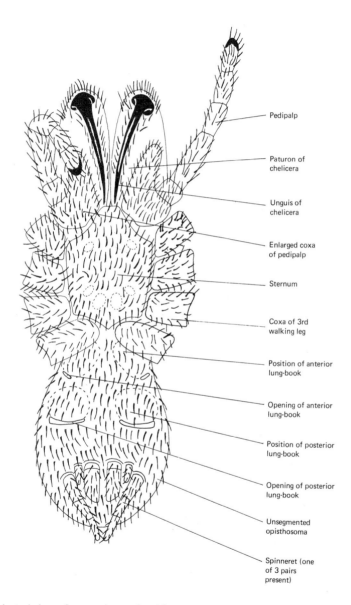

Figure 20 Ventral view of a mygalomorph spider.

Pedipalp

Paturon of chelicera

Unguis of chelicera

Enlarged coxa of pedipalp

Sternum

Coxa of 3rd walking leg

Position of anterior lung-book

Opening of anterior lung-book

Position of posterior lung-book

Opening of posterior lung-book

Unsegmented opisthosoma

Spinneret (one of 3 pairs present)

The appendages of the first segment, the chelicerae, are composed of two podomeres: a basal podomere or paturon and a distal unguis or fang. A groove which is bordered by teeth is present in the paturon, and the unguis folds back into it when at rest. Located on the convex surface of the unguis, close to its tip, is the opening of the poison gland, the gland itself being located within the paturon and extending into the prosoma. When an araneomorph spider feeds the chelicerae strike horizontally in the direction of the mid-line and the prey is pierced by the tips of the two cheliceral fangs simultaneously (Fig. 25).

The pedipalps of spiders consist of six podomeres, the general pattern being that of a walking leg except that the metatarsus is absent. In the female spider the pedipalp is clawed but in the males the distal podomere or tarsus is modified to form an organ used for the transfer of spermatozoa to the female during mating. This accessory sexual structure, known as the tarsal organ, lies within a depression or alveolus in the tarsus when not in use but is extruded during sperm transfer (Fig. 21). In its simplest form the organ consists of a proximal bulb (fundus), a reservoir and an ejaculatory duct (embolus), but the pattern may be complicated by the addition of chitinous processes which protect the organ and act as guides to aid insemination. Fig. 21 illustrates the tarsal organ of a male spider and shows one such chitinous development called the conductor.

The coxae of the pedipalps lie on either side of the mouth and are modified in both the male and female to assist in feeding. Spiders are fluid feeders and the inner border of each coxa bears several rows of setae which are used to reject large particles yet allow liquids to enter the mouth. In some species the inner

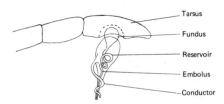

Tarsus
Fundus
Reservoir
Embolus
Conductor

Figure 21 The tarsal organ of a male spider (simplified).

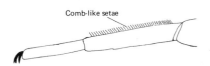

Figure 22 Ventral view of the posterior region of a cribellate spider.

Figure 23 The calamistrum (comb) on the metatarsus of the fourth walking leg of a cribellate spider.

borders of the pedipalpal coxae are armed with small denticles which help in crushing food.

Each of the eight walking legs is composed of seven podomeres, the distal podomere bearing two claws in ground spiders and three in web-spinners. A small pad is also present on the tarsus which, in web-spinners, acts as a spring and releases the claws when the spider is moving across its web. In this way the spider avoids becoming tangled in its own web.

The twelve-segmented opisthosoma usually shows no signs of external segmentation and is large and saccular. The first segment of the opisthosoma (body segment 7) forms the waist, or pedicel, and this links the prosoma with the remainder of the opisthosoma. On the ventral surface there are, primitively, a pair of lung-books on each of segments eight and nine which are seen as regions of paler cuticle. In more highly evolved spiders, however, the posterior pair are converted into spiracles with an associated tracheal system, and in many spiders including the common garden spider, *Araneus*, the two spiracles are merged together to form a single, central opening just posterior to the gonopore.

Segment eight bears the gonopore situated in the middle of a transverse groove called the epigastric furrow. On either side of the female gonopore, but still within the furrow, are openings of the spermathecae through which spermatozoa are introduced by the male.

A number of silk-spinning organs or spinnerets occur on segments ten and eleven. In some species there are four pairs of these, each spinneret being derived from a single ramus of an

D

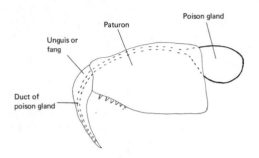

Figure 24 A chelicera and poison gland of an araneomorph spider.

originally biramous appendage. Many spiders show a reduction in the number of these organs such that only one pair remains. In a small group of spiders, known commonly as the cribellate spiders, the most anterior pair of spinnerets are replaced by a plate called the cribellum (Fig. 22).

Segments twelve to eighteen are much reduced in size and form a structure known as the anal tubercle, through the tip of which opens the anus.

2 *Mygalomorph spiders*
With a few noteworthy exceptions there is little difference between these forms and araneomorph spiders (see Figs. 19 and 20). The most obvious difference concerns the form and mode of action of the chelicerae which are large and prominent. In mygalomorphs the chelicerae strike vertically and parallel to one another such that the prey is stabbed by the tips of the two cheliceral fangs in two distinct places (Fig. 25).

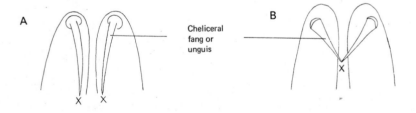

Figure 25 The form and point of contact of the cheliceral fangs of A, a mygalomorph spider and B, an araneomorph spider.

The second major difference concerns the respiratory system, for gaseous exchange is by means of two pairs of lung-books, there being no tracheal system in mygalomorphs.

Anatomy, physiology and life histories

Spiders are free-living, carnivorous arachnids which feed on insects, woodlice, diplopods, chilopods, other arachnids and, in the case of a few large tropical forms, small vertebrates. The prey is captured either by being grasped and physically overcome or by being trapped in a silken snare or web. In either case the sense organs of the spider are of great importance in detecting the prey. In diurnal ground spiders the eyes are of major importance and the prey is located by sight. However, in nocturnal ground spiders the prey is located by touch by means of body setae, and by the chemoreceptive lyriform organs on the carapace, sternum and legs. Web-spinning spiders detect the vibrations of an animal trapped in the web mainly by means of receptors on the legs.

Whether the prey is entangled in the sticky threads of a web or pounced upon it is subsequently stabbed with the cheliceral fangs

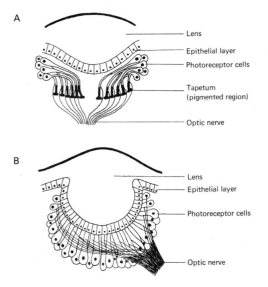

A

Lens

Epithelial layer

Photoreceptor cells

Tapetum
(pigmented region)

Optic nerve

B

Lens

Epithelial layer

Photoreceptor cells

Optic nerve

Figure 26 Vertical section through A, a lateral or indirect eye of a spider and B, an anterior median or direct eye of a spider.

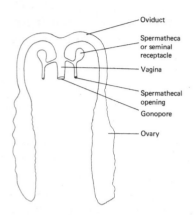

Figure 27 The reproductive system of a female spider.

and a poison issues from the opening of the poison gland (Fig. 24). In the mygalomorphs the saccular gland is restricted to the paturon but in the araneomorphs it extends into the prosoma. As the fang enters the prey the muscle fibres which surround the gland contract, so expelling a secretion which causes damage to either the nervous system or the blood of the victim, depending on the species of spider. Although there are some tropical spiders that may be harmful to man (e.g. the North American black widow, *Latrodectus* and the Australian trap-door spider, *Euctimena*), no British species is in any way poisonous.

Following the initial tearing action by the chelicerae when many of the body organs of the prey are ripped out, the prey is subjected to extra-oral digestion by a secretion from the glands found in each pedipalpal coxa which is weakly histolytic, and a secretion with strong tissue-digesting properties produced by the epithelium of the mesenteron or mid-gut. The latter fluid is rich in proteinase, peptidases, lipase and amylase and liquefies the soft parts of the prey within a few hours. The liquid food is sucked from the prey by means of the pharynx and, to a greater extent, by means of the proventriculus or 'pumping stomach'. The liquid reaches the mouth via the pre-oral food canal formed by the labrum above, the pedipalpal coxae laterally and the anterior sternum ventrally. As it passes along this canal any large particles are filtered off by the setae on the inner margins of the pedipalpal coxae.

Eventually, the food reaches the mid-gut where digestion and absorption take place. The mesenteron is extensive, filling the posterior half of the prosoma and most of the opisthosoma, and has diverticula or branches which pass into the walking legs. The posterior region of the mesenteron is modified to form a rectal sac into which open the two Malpighian tubules. Each tubule extends over the mid-gut and its diverticula and empties its guanine excretion into the rectal sac where it accumulates.

Although the Malpighian tubules are by far the most important excretory organs in spiders, the coxal glands play a part in osmoregulation. In some spiders there are two pairs of coxal glands opening on to the coxae of the first and third legs, whilst in others the posterior pair may be absent.

The products of digestion, the waste materials from metabolism and respiratory gases are transported around the body of a spider by means of a heart and circulatory system. In spiders which possess two pairs of spiracles the heart is elongate, extending over five segments and possessing five pairs of ostia. However, with the development of a tracheal system, which conducts respiratory gases directly to and from the tissues, the number of ostia and the length of the heart are reduced such that spiders which respire by trachea alone possess a heart extending over two

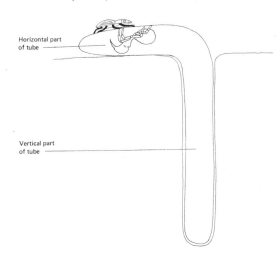

Horizontal part
of tube

Vertical part
of tube

Figure 28 Section
through the tube of the
mygalomorph spider,
Atypus, showing
the spider about to
stab an insect with
its cheliceral fangs.

segments and perforated by two pairs of ostia only. An anterior artery emerges from the heart and divides to send branches to the legs and anterior appendages and to some organs in the fore part of the body. A posterior artery leaves the heart to supply the hind regions of the gut.

Although all spiders secrete silk not all of them spin webs. The British mygalomorph spider, *Atypus*, builds a long silk tube, closed at both ends, much of which is buried vertically in the ground but with one end projecting out and lying horizontally on the surface (Fig. 28). The spider is normally located at the bottom of the tube, but when a small animal crawls over or alights on the tube the spider, on detecting the vibrations, scales the vertical tube to investigate. If the animal on the surface of the tube is a suitable prey then the cheliceral fangs penetrate the tube and the prey is paralysed by the secretion from the poison glands. After a slit has been made in the tube wall by means of the denticles on the chelicerae, the victim is pulled inside the tube. The prey is carried to the bottom of the shaft where it is left until the spider has repaired the tear in the horizontal part of the tube with silk from its spinnerets.

Other mygalomorph spiders produce a similar tube but instead of it extending over the surface it ends at ground-level where it is closed by a hinged door. Hence these araneids are referred to as 'trap-door spiders' and they wait in their tubes and pounce on passing animals. Yet other mygalomorphs do not build traps but overcome their prey by size and strength. Among these are the so-called 'tarantulas' and the bird-eating spiders.

Broadly speaking, the araneomorphs can be divided into web-spinners and ground spiders. The former group prey on flying insects which they trap in complex, two dimensional networks of silk strands covered with an adhesive. Such an arrangement is called a web. In the common garden spider the web consists of a central hub with a number of radial threads linking it with the adjacent vegetation. Immediately surrounding the hub is a spiral of about six turns which connects the radial threads. After a short gap a second spiral commences and this extends outwards for a further

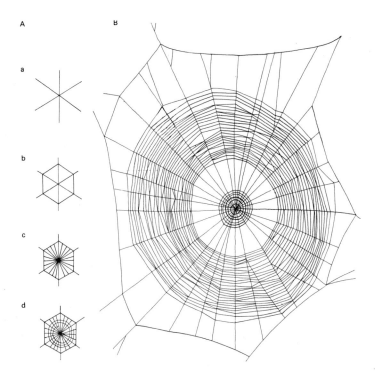

Figure 29 A. Sequence of construction of the web of the garden spider.
a. Radials constructed, b. radials joined by a spiral, c. more radials added, d. spirals completed.
B. An orb-web of the garden spider, *Araneus*.

twenty-five to thirty turns (Fig. 29). In the daytime the spider is in a retreat among the vegetation, etc. which surrounds the web but in the evening it is usually found in the centre of the web. In either set of circumstances the spider runs over the web to investigate any insect which flies into the trap, locating its position by the vibrations along the threads caused by the movements of the trapped animal. If the insect is suitable food it is immobilised and wrapped in silk thread from the spinnerets. When this has been completed the insect is carried to the centre of the web and feeding commences.

A spider's web is easily damaged and has a short existence. For this reason spiders build a new web every day, the method of construction being illustrated diagrammatically in Fig. 29.

The ground spiders hunt on or near to the ground, detecting their prey by either sight or touch. They overcome their prey either by speed and strength (Wolf spiders and Hunting spiders), or they jump on their prey from afar (Jumping spiders) or they remain hidden in vegetation, often aided by camouflage, and grasp passing insects (Crab spiders). Ground spiders do not use silk to spin webs but do spin silken retreats, and drag-lines which they secrete continuously while moving, attaching them to the substratum at intervals with an adhesive. A drag-line may be likened to the safety rope of mountaineers and prevents injury should the animal fall.

The silk glands of spiders are located within the opisthosoma and numerous ducts lead from them and open on the surfaces of the spinnerets. Silk, which chemically is a scleroprotein, is secreted as a liquid but hardens rapidly when drawn into a fine thread. As previously mentioned, a group of spiders called the cribellates has lost the anterior pair of spinnerets and replaced them by a plate, the cribellum (Fig. 22), which is perforated by a number of pores which are openings of silk glands. The silk threads which issue from the cribellar openings are combed out by a group of setae on the metatarsus of the fourth walking leg called the calamistrum (Fig. 23) and reinforce the silk from the spinnerets.

Reproduction in spiders is unique in that the tarsal organ is involved in insemination. Males of tube and web-building species leave their snares and roam freely. Each male then spins a small 'sperm web' and deposits a small quantity of semen on to it. The spermatozoa are formed in the tubular testes and are conducted to the median gonopore via the paired convoluted vasa deferentia. The male absorbs the semen into the tarsal organs and proceeds to seek a female of the same species. Courtship precedes mating and takes the form of a visual or tactile display in ground spiders, while in tube and web-building spiders the male vibrates the threads of the female's snare in a characteristic manner. The act of courtship serves three purposes: firstly, of maintaining the distinctness of species through having different courtship patterns, secondly of enabling males to be recognised as potential mates and not suitable prey, and thirdly of making females receptive by overcoming the barriers to mating. Copulation eventually takes place and is effected by the male inserting the palpal organs into the spermathecal openings (Figs. 18 and 27) of the female, either individually or both at the same time. The spermatozoa are stored in the spermathecae (seminal receptacles) and released on to the eggs as they enter the upper region of the vagina. The number of eggs laid in a single batch varies according to the species, the smallest number recorded being two while some spiders may lay several thousand. The eggs are laid on to a small web and are covered with more silk so that they are completely surrounded by an egg-sac or cocoon. This structure is then either camouflaged, guarded, or carried around by the female.

Each female usually lays several batches of eggs during her lifetime but requires only a single mating as the semen contains sufficient spermatozoa to fertilise all her eggs and the spermatozoa remain viable for long periods. Males can, in fact, inseminate several females and following each mating they spin a new sperm web and recharge their palpal organs with semen.

Within the cocoon the eggs hatch and the spiderlings emerge. After a few days they moult, at which stage they acquire eyes, spinnerets, setae, claws and the ability to feed. They are now referred to as spiders and emerge from the cocoon to be distributed over considerable distances by a method called 'ballooning'. This involves the spider climbing to the top of a blade of grass or other vegetation and secreting a long, single strand of silk which is carried into the air by the wind together with the attached spider.

Young spiders feed and require to moult in order to increase in size. Between three and ten moults are required to reach the adult stage, the actual number depending on the species. Most species live for about a year and a few even longer, the latter moulting annually to renew their exoskeletons which become damaged during the course of a year.

5

Scorpions

Man has long considered the scorpion as being an animal
warranting treatment with the utmost respect. Although
this attitude may in part be due to superstition there is an
underlying logic for these arachnids possess both powerful chelate
pedipalps with which they can deliver a painful bite and a posterior
stinging apparatus, the toxin from which produces effects varying
from minor local discomfort to paralysis and ultimate death.

Members of the order Scorpiones, to which all true scorpions
belong, are not found in Britain but are restricted to tropical and
sub-tropical countries. We are, however, ever reminded of their
existence as one genus has been adopted as a sign of the Zodiac.

External features
Scorpions are the largest of the arachnids and measure some five to
eight inches in length. The body is conspicuously divided into a
broad prosoma and a narrower opisthosoma. The latter is further
subdivided into an anterior mesosoma and a posterior metasoma
(Figs. 30 and 31).

The dorsal surface of the prosoma is covered by an unsegmented
carapace and bears a single pair of median eyes and three to five
pairs of lateral eyes. Ventrally, the surface of the prosoma is formed
from a small sternal plate and the expanded coxae of the four pairs
of walking legs. The chelicerae are small and each is composed of
three podomeres, of which the third is movable and opposes the
second to form a chelate appendage. The pedipalps are very large
and, like the chelicerae, are chelate although in this instance the
movable sixth podomere opposes the stationary fifth. Both

48

podomeres of the chela bear rows of denticles and hence the
pedipalps are formidable organs of offence and defence.
Additionally, the pedipalps function as sense organs, and are
equipped with numerous hair-like trichobothria on their surfaces,
while the pedipalpal coxae form the sides of the pre-oral food
canal.

The coxae of the four pairs of walking legs are large and are in
contact with one another such that they cover and protect much of
the undersurface of the prosoma. Two pairs of anteriorly directed
projections arise from the coxae of the first two pairs of walking
legs and together these form the lower lip of the pre-oral food
canal. All of the walking legs are clawed, each terminating in a
pair of long, curved claws situated on either side of a shorter,
median one.

The opisthosoma is narrower than the prosoma and hence its
anterior limit is easily determined. The mesosoma consists of eight
segments in the embryo but segment one is not apparent in the
adult. Each of these segments is covered dorsally by a tergal plate
and hence seven terga are visible. However, when the animal is
viewed from the ventral aspect only six sternal plates can be seen
and these are so spaced that they cover the whole of the
undersurface of the mesosoma. The seven terga and six sterna
are linked along their lateral margins by means of flexible arthrodial
membrane which, because of its location, is commonly referred to
as the pleural membrane. From embryological studies it has been
concluded that it is segment two of the embryo (adult segment one)
which lacks a sternum and that the remaining plates belong to
adult segments two to seven.

A number of modified appendages are found on the mesosoma.
Thus the first segment of the adult bears the genital operculum
which is a partially divided flap of tissue covering the gonopore.
The second adult segment bears the paired, comb-like pectines,
which are unique to scorpions, and which have been alleged to
perceive ground vibrations. The four pairs of respiratory openings
are located on the lateral margins of the sterna of the third to sixth
adult segments and each leads into an internal lung-book.

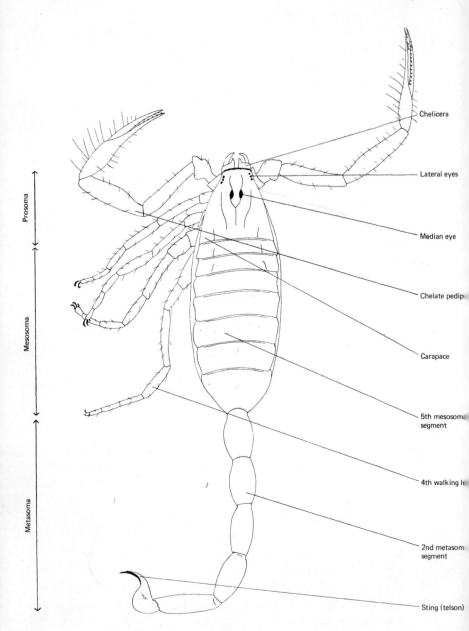

Figure 30 Dorsal view of a scorpion.

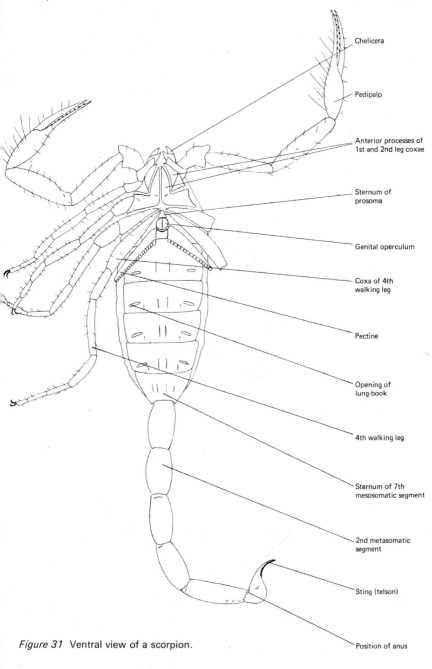

Chelicera

Pedipalp

Anterior processes of
1st and 2nd leg coxae

Sternum of
prosoma

Genital operculum

Coxa of 4th
walking leg

Pectine

Opening of
lung-book

4th walking leg

Sternum of 7th
mesosomatic segment

2nd metasomatic
segment

Sting (telson)

Position of anus

Figure 31 Ventral view of a scorpion.

The metasoma is made up of five segments and is sometimes erroneously referred to as the tail, presumably because of its narrowness and flexibility. It is almost cylindrical in cross-section and lacks appendages. The anus opens ventrally just posterior to the last metasomatic segment but it is not terminal as there follows a post-segmental telson forming the stinging organ. This possesses a sharp point which can penetrate the integument of its enemy or prey allowing ingress of venom from the gland in this region. The sting articulates with the last segment of the opisthosoma and, although its main movement is in a vertical plane, it can move laterally also (Fig. 32).

Anatomy, physiology and life histories

Scorpions are carnivorous arachnids, their main items of diet including insects, spiders and small mammals. The eyes of scorpions are incapable of forming sharp images and this, together with the nocturnal behaviour of the animals, implies that the eyes are of little use in detecting prey. Because of this the reception of air and ground vibrations becomes important in determining the position of the prey. The trichobothria situated on the surfaces of the pedipalps and the pectines are, therefore, useful in this role. The same organs are also important in the detection of approaching enemies.

The prey is grasped by means of the large chelate pedipalps and, if necessary, it is subdued by the stinging action of the telson. The poison is produced by two saccular glands which open by separate ducts near the tip of the sting. Each gland is sandwiched between a sheet of muscle and the exoskeleton of the telson so that by contraction this muscle expels the poisonous secretion. Whilst being lethal to invertebrate animals, the venom of most scorpions is harmless to man. Exceptions do occur, however, and a few species are capable of inflicting serious illness and may even cause the death of their victim. Two distinct types of toxin are produced: some scorpions secrete a poison which acts on the tissues adjacent to the site of the sting and these forms are

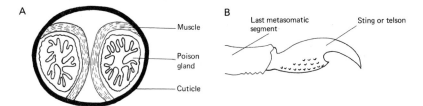

A

— Muscle

— Poison gland

— Cuticle

B

Last metasomatic segment

Sting or telson

Figure 32 Transverse section (A) and lateral view (B) of the sting of a scorpion.

harmless to man, while others inject neurotoxic and haemolytic substances causing paralysis and damage to blood cells which may result in death.

The chelae of the pedipalps crush the food, an action which is continued by the small, forwardly directed, chelate chelicerae. The fluid remains of the prey is then sucked into a pre-oral food canal formed from the labrum and the chelicerae above, the pedipalpal coxae laterally and from the anterior lobes of the coxae of the first and second walking legs below.

After the fluid food has been pumped into the alimentary canal by the action of the pharynx it passes through a short oesophagus into the long, narrow mid-gut. This region of the gut receives five or six pairs of ducts from the extensive digestive glands which occupy most of the body cavity of the mesosoma. The mid-gut leads into the short hind-gut, the two pairs of Malpighian tubules being located at their junction. These tubules constitute a part of the excretory and osmoregulatory system and are aided in the latter function by a single pair of coxal glands which open to the exterior adjacent to the base of each third walking leg.

The respiratory system of scorpions is simple and consists of four pairs of lung-books located on each of the third to sixth mesosomatic segments. There is no internal tracheal system to conduct the respiratory gases to and from the tissues and hence the heart and circulatory vessels are well developed in order to transport oxygen and carbon dioxide in the blood. The heart is long and tubular and extends the length of the seven-segmented mesosoma. In each of these segments there is a pair of ostia through which blood enters the heart and a pair of lateral arteries which conduct blood away from the heart. In addition to these emergent vessels there are anterior and posterior extensions of the heart known as the anterior and the posterior aortae.

The central nervous system retains a distinct ventral nerve cord of seven pairs of segmental ganglia, the remaining ganglia having fused with the sub- and supra-oesophageal ganglia to form the central nerve mass.

The reproductive system of the female comprises a network of ovarian tubules leading to a pair of oviducts which dilate to form the seminal receptacles before they reach the median gonopore. The male system is similar except that there are separate networks on the left and right sides, each leading by way of a vas deferens and seminal vesicle to a single, median gonopore. Reproduction in scorpions involves an extremely complex pre-mating courtship in which the animals, which are solitary normally, come together and display. The male, which is more slender and has a longer metasoma than the female, locates a mate and grasps her pedipalps with his and leads her forwards, sideways and backwards in a 'pas de deux'. In some species the metasoma and sting may also be employed in this courtship display and are held high in the air so that their stings are touching. Such behaviour may last for several hours or even days. Following successful courtship the male deposits a spermatophore on to the ground, the case of which is equipped with chitinous hooks (Fig. 33). The male then

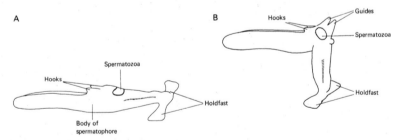

Figure 33 The spermatophore of a scorpion when deposited on the ground by the male (A), and its appearance after it has hooked on to the operculum of the female (B). As can be seen the body of the spermatophore breaks into two to expose the spermatozoa which then enter into the gonopore to effect fertilisation.

manoeuvres the female over the spermatophore which hooks on to her genital operculum and subsequently passes into her genital tract. After fertilisation the eggs develop within the female. In some species the eggs contain a large quantity of yolk and these develop without receiving nutrient materials from the female, a method of development known as ovoviviparity. In contrast to this the eggs of some species possess little yolk and most of the nutrient substances are derived from the mother. This represents a true case of viviparity.

In both cases the young scorpions climb on to the back of their mother soon after they are born and remain there for one or two weeks, by which time their first moult has taken place. During this time the female protects the young but does not feed them, their nourishment being derived from the remains of their yolk supply. Seven to eight months after leaving their parent the scorpions become adult.

E

6

Other arachnids and king-crabs

The most important arachnids have now been described in some
detail but in order to present a wider picture of the class Arachnida
a brief mention will be made of three other orders: the Pseudoscorpiones (false scorpions), the Opiliones (harvestmen) and the
Solifugae (sun spiders). In addition, consideration will be given to
the king-crabs (Xiphosura) which, although now placed in a
separate class within the Chelicerata (Fig. 2), are still thought
of as arachnids by many zoologists. Although 'crab' is used to
describe the xiphosurans it must be appreciated that this does not
imply a close relationship with the true crabs of the class
Crustacea.

False scorpions
Members of the Pseudoscorpiones are small arachnids, measuring
only two to three mm. in length. In general appearance they
resemble scorpions, although there is no sting present and no
division of the opisthosoma into a mesosoma and a metasoma.
However, they do have large chelate pedipalps, small anteriorly
directed chelicerae and general body proportions in common
with the Scorpiones (Figs. 34 and 35). Like the scorpions they
catch their prey using their toothed pedipalpal chelae but subdue
them with poison which issues from a gland in the immovable
podomere of the chela. The duct from this gland opens near the
tip of this immovable podomere and hence the poison is injected
into the victim as it is grasped. False scorpions lie in wait for
their insect or arachnid prey and when such an animal brushes
against their body it is immediately pounced upon, being detected

56

by the sensory setae on the body and appendages. As is true of most arachnids the prey is externally digested and the resultant fluid is drawn into the mouth by means of the muscular pharynx.

In common with spiders, false scorpions produce silk. This is secreted from the glands within the prosoma which open to the exterior on the movable podomere of each chelicera. In their morphology and location they are reminiscent of the poison glands of spiders and it has been suggested that they may well be homologous structures. On the distal margin of each chelicera is located a spinneret and this spins the silk for use in the construction of nests for egg-laying and for moulting.

Courtship and mating patterns are similar to those exhibited by true scorpions and entail the male grasping the pedipalps of the female with his own pedipalpal chelae and dancing around with her. The male subsequently deposits a spermatophore on the ground and then departs while the female proceeds to manoeuvre herself over the spermatophore and draw it into her genital tract. After fertilisation some two to forty eggs are laid and retained in a transparent membrane attached to the undersurface of the opisthosoma. Here the eggs eventually hatch into the larval stage and are nourished by the female with an ovarian secretion which issues from her gonopore. After moulting into protonymphs the young false scorpions become active and disperse to moult a further three times before becoming adult, passing through deutonymph and tritonymph stages. The total length of a life cycle is somewhat variable but, on average, is about a year.

Harvestmen
Most opilionids or harvestmen measure five to ten mm. in length and possess very long thin legs. The legs, together with the absence of a waist or pedicel and the presence of an externally segmented opisthosoma, permit their easy recognition and provide a difference between them and true spiders (see Figs. 36 and 37).

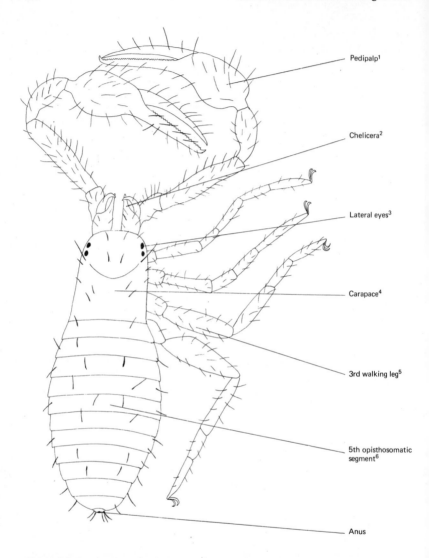

Pedipalp[1]

Chelicera[2]

Lateral eyes[3]

Carapace[4]

3rd walking leg[5]

5th opisthosomatic segment[6]

Anus

Figure 34 Dorsal view of a false scorpion. Notes on the figure: 1, The large chelate pedipalp has long, sensory setae called trichobothria. 2, The chelate chelicera is composed of two podomeres. 3, One or two pairs of eyes may by present or they may be absent. 4, The carapace covers the entire dorsal surface of the prosoma. 5, Each leg lacks a patella but the femur is divided and there are varying numbers of tarsal joints. 6, The opisthosoma is externally segmented and is composed of 12 segments.

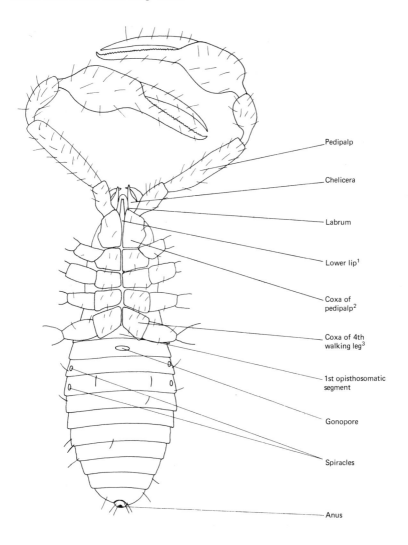

Pedipalp

Chelicera

Labrum

Lower lip[1]

Coxa of
pedipalp[2]

Coxa of 4th
walking leg[3]

1st opisthosomatic
segment

Gonopore

Spiracles

Anus

Figure 35 Ventral view of a false scorpion. Notes on the figure: 1, Formed by a forward elongation of the sternum of the prosoma. 2, The pedipalpal coxae project anteriorly and form the lateral walls of the pre-oral food canal. 3, The coxae of the legs form the skeleton of the ventral region of the prosoma, there being no visible sternum.

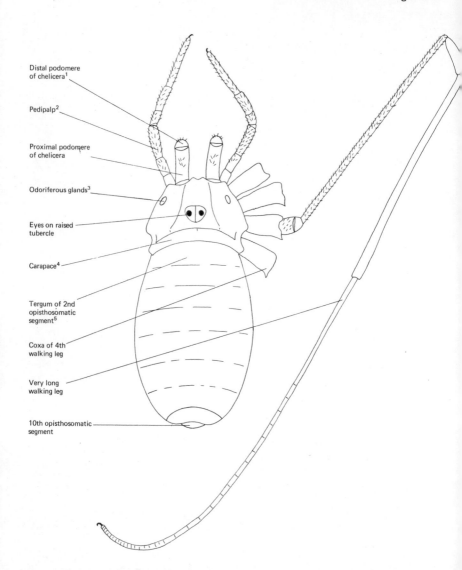

Distal podomere
of chelicera[1]

Pedipalp[2]

Proximal podomere
of chelicera

Odoriferous glands[3]

Eyes on raised
tubercle

Carapace[4]

Tergum of 2nd
opisthosomatic
segment[5]

Coxa of 4th
walking leg

Very long
walking leg

10th opisthosomatic
segment

Figure 36 Dorsal view of a harvestman. Notes on the figure: 1, The chela of the chelicera points downwards. 2, The pedipalp is leg-like and is usually clawed. 3, The odoriferous glands are used in defence but are harmless to man. 4, The carapace covers the whole of the prosoma. 5, The opisthosoma is composed of ten segments although usually only nine are visible in the adult (number one is absent).

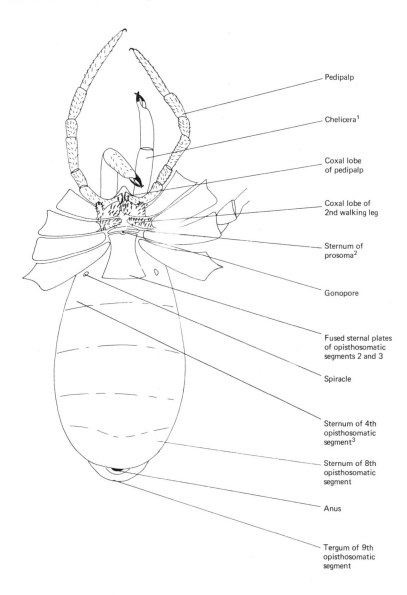

Pedipalp

Chelicera[1]

Coxal lobe
of pedipalp

Coxal lobe of
2nd walking leg

Sternum of
prosoma[2]

Gonopore

Fused sternal plates
of opisthosomatic
segments 2 and 3

Spiracle

Sternum of 4th
opisthosomatic
segment[3]

Sternum of 8th
opisthosomatic
segment

Anus

Tergum of 9th
opisthosomatic
segment

Figure 37 Ventral view of a harvestman. Notes on the figure: 1, The chelicerae
are chelate and composed of three podomeres. 2, This may be absent in some
species. 3, Only eight sternal plates are present, and two tergal plates are visible
in ventral view.

Harvestmen are omnivores and feed on living or recently dead animals, in particular small arthropods, as well as fungi and other vegetation. The pedipalps are the chief sensory organs, their setae aiding in the detection of food. However, once suitable nutrients have been located the pedipalps assist the walking legs in holding and manipulation while the food is fragmented by the chelicerae. There is no evidence of extra-oral digestion in harvestmen, and some solid food is ingested as well as fluids. In this respect opilionids are unusual amongst the Arachnida.

Each female mates several times during her life and only a few eggs are fertilised and laid at a time. Courtship is simple; the male touches the tarsi of the female with his legs and then grasps her pedipalps with his and the pair orientate their bodies so that their gonopores are opposed. The male inserts an intromittent organ (penis) into the female's gonopore and insemination occurs. Following fertilisation ten to one hundred eggs are laid in cracks and crevices or under logs and stones by means of a long ovipositor. The young hatch as miniature replicas of the adults but require seven or eight moults to become sexually mature, a process which requires some eight to nine months.

Sun spiders

The overall length of sun spiders varies between one and five centimetres, including the extremely large chelicerae which account for about one third of their total length (Figs. 38 and 39). As their morphology suggests these arachnids are voracious carnivores, their diet including insects, other arachnids, lizards, small mammals and birds. The prey is located by means of the eyes and the tactile setae on the first pair of legs. These appendages are slender compared with the other walking legs and are held out in front of the animal and take no part in locomotion. The prey is killed, not by poison, but mechanically by the formidable chelicerae which proceed to crush the prey, the resultant fluids being drawn into the mouth by the action of the pharynx. As in harvestmen, it appears that there is no phase of extra-oral digestion

in sun spiders. Prior to mating the male courts the female by
caressing her with his appendages, and proceeds to turn her on
to her side and release spermatozoa on to the ground. He picks
up the semen with his chelicerae, transfers it into the gonopore
of the female and then retires before the female returns to her
normal agressive state. Subsequently the female constructs a
burrow within which she lays her one hundred to two hundred and
fifty eggs. These eventually hatch into larvae and several moults
follow before the adult stage is reached, the precise number
having not yet been determined.

King-crabs

Unlike the animals previously described the king-crabs (Figs. 40
and 41) are marine animals and are restricted to shallow
waters off the eastern coasts of South America and Mexico,
and the shores of the Middle and Far East.

King-crabs are scavengers and plough through the mud or sand
of the sea-bottom in search of molluscs, polychaete worms and
algae. When not searching for food they remain stationary, partially
buried and well protected from their predators. Food is grasped
with the chelicerae and passed to the gnathobases of the pedipalps
and first three pairs of walking legs where it is triturated before
entering the mouth. In this action they differ from the true
arachnids in that solid particles are ingested as the *main* item
of diet.

Mating takes place in very shallow water and commences with
the male climbing on to the back of the female, using his hooked
pedipalpal tarsi in order to retain a firm hold. The female responds
by excavating a small hole in the substratum in which she
deposits some one hundred and fifty to three hundred eggs.
Immediately after the female has completed oviposition the male
sheds spermatozoa into the water in the vicinity of the eggs and
fertilisation occurs when spermatozoa and eggs come into contact.
The pair subsequently separate, leaving the female to cover the
eggs with sand or mud. After some time the eggs hatch to give the

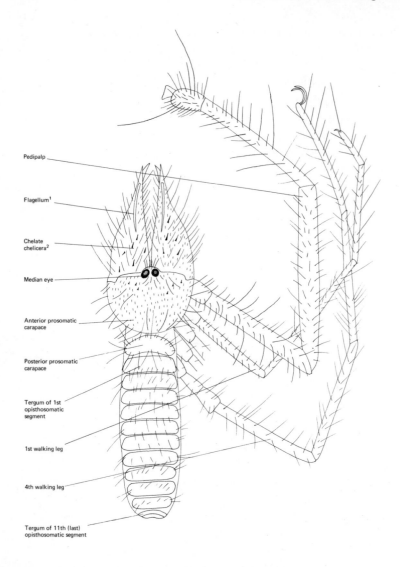

Pedipalp

Flagellum[1]

Chelate
chelicera[2]

Median eye

Anterior prosomatic
carapace

Posterior prosomatic
carapace

Tergum of 1st
opisthosomatic
segment

1st walking leg

4th walking leg

Tergum of 11th (last)
opisthosomatic segment

Figure 38 Dorsal view of a sun spider. Most of the setae have been removed from
the opisthosoma. Notes on the figure: 1, This organ is present in the male only
and is of questionable function. 2, The chelicera is enormous and composed of two
podomeres.

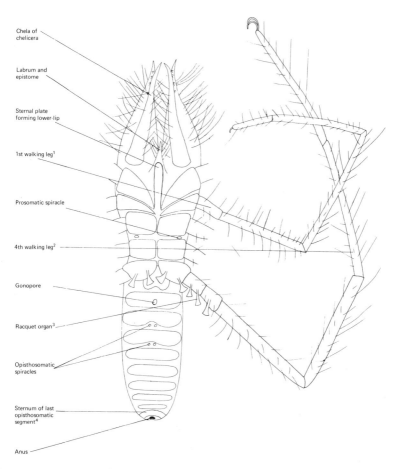

Figure 39 Ventral view of a sun spider. All the setae have been removed from
the opisthosoma and the coxae. Notes on the figure: 1, The first walking leg is thin
and is used in feeding and as a sense organ. 2, The second, third and fourth legs
are stout and used in locomotion. 3, Two sensory racquet organs are present on
the coxa and three on the divided trochanter of the fourth walking leg. 4, This
sternum is united to the last tergum forming a perianal ring.

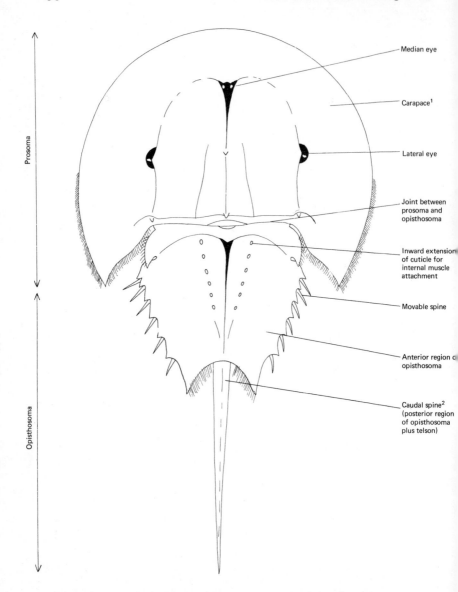

Figure 40 Dorsal view of a king-crab. Notes on the figure: 1, Horse-shoe shaped carapace covering the prosoma. 2, The caudal spine is used for righting the animal if turned over and for locomotion (punting). It is composed of the 8th to 10th segments of the opisthosoma and the telson.

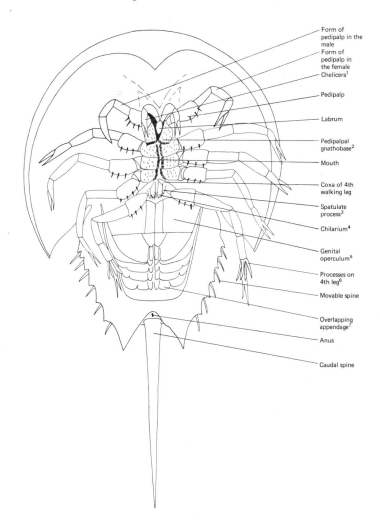

Form of pedipalp in the male
Form of pedipalp in the female
Chelicera[1]
Pedipalp
Labrum
Pedipalpal gnathobase[2]
Mouth
Coxa of 4th walking leg
Spatulate process[3]
Chilarium[4]
Genital operculum[5]
Processes on 4th leg[6]
Movable spine
Overlapping appendage[7]
Anus
Caudal spine

Figure 41 Ventral view of a king-crab. Notes on the figure: 1, Each chelicera is small and composed of three podomeres. 2, The bases of the pedipalps and the first three pairs of walking legs form crushing areas called gnathobases. 3, Used for cleaning the gills. 4, These are the sensory appendages of the 1st opisthosomatic segment. 5, The genital operculum covers the two gonopores and is composed of the fused appendages of the second opisthosomatic segments. 6, These processes are used for digging and for walking on soft mud. 7, These appendages belong to the 3rd to 7th opisthosomatic segments and bear gills on their undersurfaces.

free-swimming, dispersive larval stage, which resembles the adult closely except that it lacks a caudal spine. After several moults the spine is fully developed and the animal has all of the adult characteristics. When fully grown king-crabs may reach a length of sixty cm. although ten to twenty cm. is more usual.

The form of fertilisation found in king-crabs differs from that of true arachnids in that it occurs outside the body of the female. The majority of aquatic invertebrates exhibit external fertilisation for this is suitable for animals living in a watery medium. However, terrestrial animals cannot employ this method and have developed internal fertilisation so that the spermatozoa and unfertilised eggs are not subjected to an unfavourable environment. Following internal fertilisation eggs are provided with impermeable shells and so are shielded from adverse conditions.

7

Collecting, preserving and studying arachnids

Before an attempt is made to collect examples of any group of animals it is necessary to have a knowledge of their ecology, life-histories and habits. Thus the collector not only searches in the correct geographical area but confines his investigation to the precise microhabitat which is occupied by the particular animal. In addition to this he will also know the time of year or day at which to search and will recognise all stages in the life-cycle of the animal and not merely the adult form.

Scorpions are the largest of the arachnids and inhabit tropical and sub-tropical lands. In Europe they are found in Spain, Italy, Greece, the Balkans and Southern Germany, and also occur throughout Southern Asia, Africa, Australasia, the southern half of the United States and throughout most of South America. They are nocturnal animals and, during the daytime, are found under rocks, stones or bark or in burrows which they construct for themselves. Care should be taken when dealing with these arachnids as they may be poisonous and even if not they have formidable pedipalps which may cause a painful bite. It is, therefore, advisable to manipulate them with long, stout forceps.

Sun spiders are also tropical and sub-tropical in their distribution and occur throughout Africa, Southern Asia as far east as India, in Indo-China and Celebes, along the western coast of South America and the United States, and in South-East Spain. Again, sun spiders are nocturnal and may be found roaming freely in desert regions at night or under debris, stones, etc. during the daytime.

In contrast to the restricted distribution of scorpions and sun spiders, false scorpions, spiders and mites are ubiquitous.

The majority of species of false scorpions are found in the soil and in decaying vegetation and may be collected by sifting samples of soil or leaf litter in a white dish. Small spiders, free-living mites and unattached parasitic mites and ticks may also be found by this procedure but larger spiders are best sought under stones, on vegetation, on or near webs or in their tubes. When located they may usually be caught in a net, although in the case of tube-dwellers the only satisfactory method of capture is to excavate the whole tube and then remove the occupant.

Ticks and parasitic mites may be collected by removing them from their host. This presents difficulties for their hosts, which are mammals, birds and reptiles, are not easily located and even so the chances of finding a parasitised host are often small. Another consideration is, of course, that permission may be necessary to trap vertebrate hosts and this may not be given readily. Unattached ticks and parasitic mites may be collected by dragging a blanket over the ground. The blanket is examined at intervals to remove the acarines which have adhered to it.

In order to preserve arachnids for the future investigation of their external features they may be placed in 70% alcohol. Alcohol of this concentration is convenient to use as it does not stiffen muscles and hence the appendages may be moved freely during examination of external morphology. It does have limitations in that specimens become brittle if allowed to dry out during investigation and it causes loss of pigment to occur in coloured specimens, but with regard to the arachnids this latter point is not too serious, at least as far as routine studies are concerned.

When specimens are collected in the field they should be placed in a container of 70% alcohol together with a label noting the date and place of capture. When the animal is later identified by the use of guides and keys the scientific and common names should be added to the original label.

For the purpose of examining arachnids, whether for identification or for studying the external features, the animal is viewed under a binocular microscope. Although it is not advisable to permit specimens to dry out it may be found helpful to blot their surface or allow the surface to dry in order to see the features more clearly. It is often useful to place the specimen on the top of a wedge of 'Plasticine' modelling clay for then the orientation of the specimen may be changed, thus permitting a more thorough examination of its morphology. When dealing with small arachnids, for example false scorpions, or when investigating the detailed morphology of a particular region of a larger arachnid, for example the hypostome of a tick, the specimen must be viewed beneath a high-power microscope. Before it can be examined the specimen requires to be processed. Many different methods have been devised to prepare specimens for examination but the following is simple and produces a permanent slide:

1 Place fresh specimen in 70% alcohol for at least 10 minutes.

2 Dehydrate the specimen by passing it through 90% and absolute (100%) alcohols. Again at least 10 minutes should be spent in each solution.

3 Transfer specimen to xylene for about 10 minutes in order to 'clear' the tissues, that is make them transparent.

4 Place a drop of Canada balsam on to a glass slide and transfer the specimen into the balsam. Lower a cover-slip over the specimen.

Alternatively, a temporary mount may be made by transferring the specimen from 70% alcohol to lactic acid, allowing it to remain in this medium for several hours, and mounting it under a cover-slip, again in lactic acid. The process may be hastened by mounting the specimen in lactic acid within minutes of its initial transference and gently warming the slide over a low flame until the specimen clears.

The internal anatomy of arachnids may be investigated by dissection or sectioning. Large animals such as scorpions can be dissected by pinning them down in a dish containing solidified wax and using normal dissecting equipment. Smaller animals, for example spiders and ticks, may be anchored to a similar dish by melting the wax and pressing their appendages into it. When the wax solidifies the animal is held securely and is then ready for

F

dissection beneath a binocular microscope. Normal dissecting instruments are too large and special tools must be used. These can be purchased, although cutting tools are easily constructed by setting slithers of razor blades or fine entomological pins into glass or wooden shafts.

Section cutting presents difficulties as the cuticle of arachnids is hard and does not permit the rapid entry of the processing chemicals. The problem of hardness can be overcome to some extent by using animals which have just hatched or moulted for at this time their exoskeletons are soft. However, internal changes often occur at the time of moulting and hence incorrect deductions may be made from such a method of investigation.

An acceptable method of processing arachnids for sectioning is outlined below. Suggested times for each step are given but these will vary for different animals and even the same animal of different size.

1 Immerse the living animal in Petrunkevitch's fixative for 2 to 5 days. This fixative has the following composition:

70% alcohol	100 ml
ether	5 ml
concentrated nitric acid	3 ml
para-nitrophenol	5 g
cupric nitrate	2g

2 Wash out the fixative with several changes of 70% alcohol.

3 Dehydrate by passing through 90% and absolute alcohols. Leave the specimen in each solution for 12 to 24 hours.

4 Transfer the specimen to a solution of equal parts by volume of absolute alcohol and ether for 1 day.

5 Place the specimen in a 2 to 3% solution of celloidin in absolute alcohol/ether for 1 to 2 days.

6 Harden the celloidin and clear in chloroform for about 12 hours.

7 Transfer the specimen to a high melting point moulten wax (58°C M.P. is suitable), and change the wax every two hours. The total

time required for the wax to penetrate the animal will vary and must be decided by experiment but 12 hours is an average time.

8 Embed the specimen; cut sections at 8 to 10μ; stain in either Mallory's triple stain or haematoxylin and eosin.

The instructions given above assume a prior knowledge of certain histological techniques and merely show how the basic methods can be modified for use with arachnids. A number of books are recommended in the reading list and these should serve for the beginner as well as the more experienced student.

8

Glossary

Ambulatory Concerned with walking.

Anatomy The study of the internal structure of organisms.

Anticoagulant Substance which prevents or delays the clotting of blood.

Anterior The front of an animal; towards the head region.

Anus The posterior opening of the alimentary canal.

Appendage Limb or projection, usually paired, arising from the body surface.

Arthrodial membrane Flexible, chitinous membrane found between the hard, rigid segmental plates of the arthropod exoskeleton.

Aquatic Living in the water.

Bilaterally symmetrical Able to be divided into two equal portions along the antero-posterior axis such that two mirror images are formed.

Blastocoele Cavity which appears in the egg during embryonic development.

Carapace The hard plate which covers all or part of the dorsal surface of some arachnids.

Carnivorous Adjective describing animals which feed on other animals, and which kill their victim (prey) prior to ingestion.

Caudal Describing the tail or tail-like structure; in the direction of or closer to the tail.

Chelicera The appendage of the first segment in arachnids.

Chemoreceptor Sense organ which perceives certain dissolved or air-borne chemicals.

Chitin A polysaccharide found in the exoskeleton of arthropods. Chemically it consists of large numbers of glucosamine units linked together.

74

Coelom The mesoderm lined cavity which normally surrounds the gut and body organs.

Copulation The mating act whereby the male introduces spermatozoa or a spermatophore into the female.

Courtship The period before copulation during which the male and female take part in ritualised movements, postures etc. in order to lower the resistance to mating. Each species has a distinct courtship pattern which prevents inter-specific mating.

Coxa The most proximal podomere of the arthropodan appendage.

Cuticle The outer covering of arthropods which makes up the exoskeleton.

Dermal Describing the skin layer beneath the epidermis.

Digestion Process whereby complex food materials are converted into simple substances which may then be absorbed by the animal.

Distal Situated away from the point of origin of an organ.

Diurnal Adjective describing animals which are active during the day and which rest at night.

Diverticulum A blindly ending region of the gut.

Dorsal Pertaining to the back of an animal, which in arachnids is the uppermost surface and is directed away from the substratum.

Ecdysis The periodic shedding of the exoskeleton by arthropods. Also called moulting.

Ecology The study of the inter-relationships between organisms and their environment.

Ectoparasite An animal which attaches itself to the external surface of another animal in order to feed from that animal.

Endoparasite An animal which lives inside another animal (its host) and which obtains its nutriment at the expense of that animal. Endoparasites are harmful to their hosts.

Evolution The process by which organisms have changed throughout time from ancestral to modern forms.

Excretion The process by which waste products of metabolism are removed from an organism.

External Digestion The digestion of food material outside the alimentary canal before it is ingested.

Extra-oral Digestion See external digestion.

Fertilisation The fusion of male and female gametes.

Free-living Describing an animal which does not live on or in another animal i.e. is not a parasite, a symbiont or a commensal.

Gamete A mature reproductive cell of an animal. A male gamete is called a spermatozoon and that of the female is known as an ovum.

Ganglion Segmentally arranged collection of nerve cell bodies which make up the nervous system of arthropods.

Gill An organ which is found in aquatic animals and which is employed in gaseous exchange.

Gonad The organ of animals which produces gametes.

Gonopore The opening into the genital system.

Guanine The nitrogenous excretory product of arachnids.

Haemocoel The specialised body cavity of arthropods which serves as a blood sinus.

Hexapod Having six legs.

Histolytic Possessing the property of causing tissue lysis.

Homologous A term describing organs in different animals which may have the same or different functions but have a similar development from a similar embryonic tissue.

Hydrostatic Skeleton Skeleton of certain animals, including annelids, consisting of a liquid under pressure contained within the body cavity.

Ingestion Process by which food is taken into the alimentary canal.

Insemination The transference of spermatozoa from the male into the female.

Instar Stage of development of arthropods between two ecdyses.

Labrum The upper lip of arthropods.

Larva The first immature stage of many animals including arthropods which hatches from the egg and in arachnids moults into a nymph or pre-adult instar.

Lateral Pertaining to the side of an animal or towards the side.

Lumen The cavity within a hollow organ, especially that of the gut.

Malpighian Tubule Excretory tubule of many arthropods arising from the mid-gut/hind-gut boundary and extending into the haemocoel.

Marine Living in the sea.

Mastication Mechanical breaking down of food by means of the mouthparts or associated structures.

Metabolism The chemical reactions taking place within an animal by which complex chemicals are converted into simple ones with the release of energy which is used for synthesis and the activities of the animal.

Metamerically Segmented The serial repetition of organs along the longitudinal axis of the animal.

Morphology The study of the external structure of an organism.

Moulting See ecdysis.

Nocturnal Being active at night and resting by day.

Nymph The immature stage which moults from the larva and which preceeds the adult.

Oesophagus That part of the alimentary canal between the pharynx and the mid-gut.

Omnivore An animal which feeds on both animals and plants.

Osmoregulation Process by which animals control the concentration and composition of the solutes of the body fluids.

Ovary The gonad of the female which produces the female germ cells or ova.

Oviduct The tube leading from the ovary toward the exterior.

Oviparous Describing a species which lays eggs.

Oviposition The process of egg-laying.

Ovoviviparity The retention and development of eggs within the female genital tract. The nutrient materials are stored within the eggs as yolk and are not transferred to the young from the mother.

Parasite An animal which lives on or in another animal and which feeds at the expense of that animal.

Pharynx That part of the alimentary canal between the oral cavity and the oesophagus.

Phylogeny The evolutionary history of an organism.

Phylum A primary division of the animal kingdom.

Physiology The study of the processes which take place within animal tissues.

Posterior The hind region of an animal or nearer to the hind region.

Predator An animal which preys upon other animals.

Pre-oral Food Canal Channel anterior to the mouth through which fluid is drawn into the alimentary canal.

Prey An animal which is eaten by a predator.

Proctodaeum An ectodermal intucking forming the posterior region of the gut.

Proximal Close to the point of origin of an organ.

Respiration The exchange of oxygen and carbon dioxide between an organism and its environment.

Rickettsia Micro-organisms which have characters in common with both viruses and bacteria.

Salivary Glands Paired glands which open into the anterior region of the alimentary canal and which produce a secretion to aid feeding.

Scavenger An animal which feeds on the remains of animals and plants.

Secretion A fluid which is produced by a gland or tissue and which has a useful function.

Sense Organ A specialised organ for the perception of stimuli, usually from the external environment.

Species A natural population of organisms which interbreed freely to produce viable offspring and which resemble one-another constantly in morphological and physiological features.

Spermatophore A sac-like structure containing a large number of spermatozoa.

Spermatozoon A male reproductive cell.

Spiracle The external opening of the arthropod tracheal system.

Stomodaeum An ectodermal intucking forming the anterior region of the gut.

Substratum Surface on which an animal stands and moves, or to which it is attached.

Suctorial Having a sucking action.

Tactile Concerned with the sense of touch.

Tagma A region of the arthropod body which is composed of a number of segments and which is distinct from the adjacent region, e.g. prosoma, opisthosoma.

Taxonomy The study of the classification of animals and other organisms.

Terrestrial Living on land.

Testis The male gonad which produces the male gametes or spermatozoa.

Thermoreceptor Sense organ concerned with the detection of temperature changes.

Trachea Tube leading from the spiracle to a body organ which conducts respiratory gases to and from the tissues.

Translocation Movement of soluble substances through the body.

Triploblastic Possessing a body composed of three embryonic layers viz. ectoderm, endoderm and mesoderm.

Triturate To reduce in size mechanically, i.e. crush.

Unsegmented Not composed of segments or not visibly segmented.

Vas Deferens Tube which conveys spermatozoa from the testis to the exterior.

Vector An animal which transmits parasites.

Ventral That surface which in arachnids is the underside and is directed towards the substratum. Opposite of dorsal.

Viviparous The development of embryos within the mother whereby the young are nourished by the mother and are born as miniatures of the adults.

Yolk The fat and protein store of eggs which provides the embryo with nutrients.

9

Further reading

The following list of publications is intended as a guide to enable the reader to persue the various subjects further. In turn these texts will direct the student to more advanced and detailed literature, including original research publications.

Arachnids in general

Cloudsley-Thompson, J. L. (1968). *Spiders, Scorpions, Centipedes and Mites.* Pergamon Press. London.

Grassé, P-P. (ed.) (1949). *Traité de Zoologie, Anatomie, Systématique, Biologie.* Volume 6. Masson et Cie., Paris.

Savory, T. H. (1942). *The Spider's Web.* Warne. London & New York.

Savory, T. H. (1964). *Arachnida.* Academic Press. London & New York.

Snodgrass, R. E. (1948). The Feeding Organs of Arachnida, including Mites and Ticks. *Smithsonian Miscellaneous Collections* *110* (10), 1–93.

Mites and ticks

Arthur, D. R. (1952). Economic Importance of Ticks. *Discovery* *13*, 379–83.

Arthur, D. R. (1962). *Tricks and Disease.* Pergamon Press. London.

Arthur, D. R. (1963). *British Ticks.* Butterworths. London.

Baker, E. W. & Wharton, G. W. (1952). *An Introduction to Acarology.* Macmillan. New York.

Hughes, T. E. (1959). *Mites, or the Acari.* Athlone Press. London.

Spiders

Bristowe, W. S. (1939—41). *The Comity of Spiders*. 2 Volumes. Ray Society. London.

Bristowe, W. S. (1954). The Chelicerae of Spiders. *Endeavour* *13*, 42—9.

Bristowe, W. S. (1958). *The World of Spiders*. Collins. London.

Crompton, J. (1954). *The Life of the Spider*. Mentor (New American Library). New York.

Locket, G. H. & Millidge, A. F. (1952—3). *British Spiders*. 2 Volumes. Ray Society, London.

Savory, T. H. (1928). *The Biology of Spiders*. Sidgwick and Jackson. London.

Scorpions

Cloudsley-Thompson, J. L. (1955). Some Aspects of the Biology of Centipedes and Scorpions. *Naturalist* (1955), 147—53.

Vachon, M. (1952). *Études sur les Scorpions*. Institut Pasteur D'Algérie. Algér.

Vachon, M. (1953). The Biology of Scorpions. *Endeavour 12*, 80—9.

Practical work

Knudsen, J. W. (1966). *Biological Techniques. Collecting, Preserving, and Illustrating Plants and Animals*. Harper and Row. New York, Evanston and London.

Mahoney, R. (1966). *Laboratory Techniques in Zoology*. Butterworths. London.

McManus, J. F. A. & Mowry, R. W. (1964). *Staining Methods*. Harper and Row. New York, Evanston and London.

Pantin, C. F. A. (1964). *Notes on Microscopical Techniques for Zoologists*. Cambridge University Press.

Index

Page numbers in italic indicate illustrations on those pages